MULTIOBJECTIVE OPTIMIZATION IN WATER RESOURCES SYSTEMS

The Surrogate Worth Trade-off Method

DEVELOPMENTS IN WATER SCIENCE, 3

advisory editor
VEN TE CHOW
Professor of Hydraulic Engineering
Hydrosystems Laboratory
Civil Engineering Building
University of Illinois
Urbana, Ill., U.S.A.

MULTIOBJECTIVE OPTIMIZATION IN WATER RESOURCES SYSTEMS

The Surrogate Worth Trade-off Method

YACOV Y. HAIMES
Systems Engineering Department
Case Institute of Technology
Case Western Reserve University
Cleveland, Ohio

WARREN A. HALL
Civil Engineering Department
Colorado State University
Fort Collins, Colorado

HERBERT T. FREEDMAN
Systems Engineering Department
University of Pennsylvania
Philadelphia, Pennsylvania

ELSEVIER SCIENTIFIC PUBLISHING COMPANY
AMSTERDAM — OXEORD — NEW YORK 1975

ELSEVIER SCIENTIFIC PUBLISHING COMPANY
335 Jan van Galenstraat
P.O. Box 211, Amsterdam, The Netherlands

AMERICAN ELSEVIER PUBLISHING COMPANY, INC.
52 Vanderbilt Avenue
New York, New York 10017

ISBN: 0-444-41313-8

Printed in The Netherlands

Preface

Large scale systems, and in particular water resources systems, create special problems which make the application of optimization methodologies quite difficult and unless treated with considerable insight quite meaningless if not actually misleading. Most of these difficulties stem from three important characteristics of these systems. First, there is a large number of quasi-independent decision-makers and/or constituencies , each of which may make decisions or influence decisions according to his own, but different, versions of the desired goals. Second, even for any one decision maker, there is a large number of non-commensurable objectives to be optimized. Third, there is a very large element of uncertainty and risk in virtually all water resources decisions. This element is due to the high degree of irreversibility of these decisions coupled with both hydrologic uncertainty and inability to predict the future with reasonable accuracy.

The recent trend in water resources and other real world problems, however, has been to elevate many of the non-commensurable objectives to parity with economic efficiency as criteria for excellence. This inclusion of a vector of objective functions introduces a new dimension in the fields of modeling, mathematical programming and optimal control. This book represents a comprehensive survey of the methods presently available for solving multiple objective problems, and concentrates on a new powerful and operational method, namely the Surrogate Worth Trade-off (SWT) Method. This method simplifies the interaction between decision-makers and systems analysts, and enables them to determine a best policy via a very moderate interaction. The method is applicable to static (mathematical programming) and dynamic (optimal control) systems. Theoretical bases as well as detailed computational algorithms are discussed with several example problems. The incorporation of special properties of water resources systems (e.g. risk and uncertainty) into a multiobjective framework is analyzed and applications of the SWT method are presented. In particular, sensitivity, irreversibility and optimality are studied in detail as multiple objective functions. The availability of operational methodologies such as the SWT method, encourages and enhances the systems modeling and pattern of thinking in multiobjective function terms. Thus, more realistic analyses may result by eliminating the needs for a single objective formulation.

This book should serve college students and professors , practical engineers, and managers involved in the decision-making process of real world problems whether in water resources systems or other large scale systems. Possible areas of application include transportation, health care delivery, communication, urban and housing development, environmental and energy problems and many other areas, where noncommensurable objective functions dominate these problems.

November 1974

Yacov Y. Haimes

Warren A. Hall

Herbert T. Freedman

Acknowledgments

The authors wish to thank all individuals who have contributed to this book by their comments, critique, and suggestions. Among these individuals are Professor L. S. Lasdon (Case Western Reserve University), Professor David Marks (MIT), Professor Jared Cohon (Johns Hopkins University), Professor Charles Howe (University of Colorado), and Dr. W. Scott Nainis (Arthur D. Little, Inc.). Special thanks are due to Prasanta Das who has diligently proofread the final manuscript and to Sue Reeves who typed the manuscript. Last and not least, we thank all the graduate students in the Water Resources Program, Systems Engineering Department, at Case Western Reserve University, who have offered many comments and suggestions during the preparation of this book. The cooperation of the American Geophysical Union, and the American Society of Civil Engineers, by releasing copyrighted material to be included in the book, is appreciated.

The preparation of the material in this book was partially supported by the National Science Foundation, Research Applied to National Needs Program, under research project: "Multilevel Approach for Regional Water Resource Planning and Management", Grant Number GI-34026. Special thanks are due to Dr. Richard Kolf and to Dr. Lawrence Tombaugh who served as project officers.

TABLE OF CONTENTS

LIST OF FIGURES

LIST OF TABLES

Chapter 1

FUNDAMENTALS IN MULTIPLE OBJECTIVE PROBLEMS

1.1   INTRODUCTION

A recent trend in systems analysis has been the consideration of problems which have more than one objective function. This is especially important in the study of large scale systems, where there tend to be several conflicting and non-commensurable objectives that the system modeler can identify. For example, in water resources planning, one wants to maximize both economic efficiency, which is measured in monetary units, and environmental quality, which is measured in units of pollutant concentration. Traditionally, only one objective (economic efficiency) was considered, with the other objectives being included as constraints, or somehow made commensurate with the primary objective. However, society is placing an increasing importance on non-pecuniary objectives which are difficult to quantify monetarily. Multiple objective analysis has been applied to a wide variety of problems including transportation, project selection for research activities, economic production, the quality of life, managing an academic department, game theory, and many others.

A fundamental characteristic of decision processes is the development of logical bases for eliminating from further consideration large numbers of otherwise possible decisions, with reasonable assurance that the most desirable decision is not inadvertently lost. The reduced field of possibilities can then be more easily analyzed by a decision maker in order to arrive at a final decision.

If two or more objectives are not commensurable, then there is generally no single optimum decision. Despite this assertion, decisions which involve apparently non-commensurable objectives are reached every day by millions of people. A substantial percentage of these individuals are quite sure they made the best decision -- best in the sense that no other could be demonstrated as superior. Thus it would appear that the problem is one of finding the means of reducing non-commensurable objectives to an appropriate common denominator.

Much of pricing theory in economics is devoted to this question. Physically non-commensurable quantities which are traded in large numbers in a "free," non-coercive market, appear to have been rather well commensurated in monetary units. This has encouraged development of strategies to create, by law, the institutional equivalent of a market for the remaining non-commensurable objectives of water resources. "Pollution certificates, effluent charges, scarcity based pricing," etc. are examples of this approach.[1]   While very attractive in some respects, it is clear from other institutionally managed markets that these economic artifices in many ways may be far from adequate to create pseudo-market conditions which would in fact represent even the important objectives to any satisfactory degree.

The reluctance of the political system to adopt such pseudo-market institutions, and the acceptance of direct political allocation suggests that for the immediate future, at least, it will be necessary to seek other alternatives for treating the non-commensurable objective problem.

Thus the development of mathematical techniques for the solution of multiple objective problems is quite important. The purpose of this book is to investigate computational procedures for the solution of multiple objective problems concentrating on the surrogate worth tradeoff (SWT) method.[2]

This chapter will present the mathematical formulation of the general multiple objective problem, and discuss the concepts and terminology inherent to such problems.

1.2   MULTIOBJECTIVES IN WATER RESOURCES SYSTEMS

Water resources systems create special problems which make the application of classical optimization methodologies quite difficult and, unless treated with considerable insight, quite meaningless if not actually misleading. Most of these difficulties stem from three important characteristics of these systems. First, there is a large number of quasi-independent decision makers and/or constituencies, each of which may make or influence decisions according to their own, possibly different, versions of the desired goals. Second, even for any one decision maker, there is a large number of non-commensurable objectives to be optimized. Third, there is a very large element of uncertainty and risk in virtually all water resources decisions. This element is due to the high degree of irreversibility of these decisions coupled with both hydrologic uncertainty and inability to predict the future with reasonable accuracy.

So long as one objective (e.g. economic efficiency) dominates over all others and a single point of view (e.g. national) can be asserted as primary, the optimization can proceed along classical lines using either judgment or mathematical decision models as desired, where secondary objectives and points of view can be taken into account through judgment-based constraints. To a limited extent the judgement-based constraints can be parameterized and/or subjected to a sensitivity analysis.

The recent trend in water resources, however, has been to elevate many of the non-commensurable objectives to parity with economic efficiency as criteria for excellence.

Water resource projects are generally constructed to serve multiobjectives. This fact is inherent in the nature of almost any large-scale project, e.g., reservoirs, dams, aqueducts, the development of groundwater systems, and so on. A large reservoir created by a high dam may supply water for irrigation, municipal and industrial needs, provide for fishing and recreation facilities, improve navigation and flood control capabilities, generate hydroelectric power, maintain suitable water quality for both ground and surface water, provide a buffer for drought years and groundwater recharge, improve related land use and prevent damages from runoff, and enhance the regional development in terms of a better economy and quality of life. In regional planning of water and related land resources, the simultaneous consideration of more than one project is often essential due to the interactions and coupling that exist among them.[3] Clearly, the problem of multiprojects -- multiobjectives planning becomes truly large scale and complex. Probably one of the major reasons for the relative scarcity of multiobjective formulations and considerations in the literature, not necessarily limited to water resources systems, is that until recently, almost all the solution strategies developed involved a single objective function. Optimization techniques and methodologies are viewed here as solution strategies that are applied to the mathematical model defined by an objective function and a set of constraints. The inclusion of a vector of objective functions introduces a new dimension in the fields of modeling, mathematical programming, and optimal control, especially since the numerical notion of an optimal solution generally will not exist for a vector optimization problem, as will be discussed later.[4] It is important to note, however, that both judgment and mathematical models of decision processes have one common feature in that they utilize a logical argument to eliminate large numbers of possible decision sets from further contention for the "best" decision.

Some procedures which can accomplish this in the multiple-objective context will be explored in this book.

A number of studies have been conducted which include multiple objectives in water resources planning. The Corps of Engineers[5] used three objectives (national income, regional development, and environmental quality) in their study of the North Atlantic region of the United States. Miller and Byers[6] studied the tradeoff between environmental quality and income for an agricultural area. Cohon and Marks[7] evaluated the tradeoff between net national income and equity of regional income distribution for a developing country. Major[8] took regional development into account in traditional cost-benefit analysis. O'Riordan[9] used the objectives of economic growth, environmental quality **and** social well-being for river basin planning in Canada. Monarchi et al[10] presented a sequential technique which should enable the decision maker to determine a satisfactory solution from non-inferior points. An analysis of the applicability of various multiple objective techniques to water resources problems has been carried out by Cohon.[11]

1.3   PROBLEM DEFINITION

For notational convenience, define the general vector optimization to be:

Problem 1-1:

$$\min_{\underline{x}} \{f_1(\underline{x}), f_2(\underline{x}), \ldots, f_n(\underline{x})\}$$

Subject to

$$g_k(\underline{x}) \leq 0 \quad , \quad k = 1, 2, \ldots, m$$

Where $\underline{x}$ is an N - dimensional vector of decision variables.

$f_i(\underline{x})$, i = 1,2, ..., n, are n objective functions.

$g_k(\underline{x})$, k = 1,2, ..., m, are m constraint functions.

For simplicity in notation, equality constraints are not present. It can be assumed that each equality constraint was replaced by two inequality constraints (≥ 0 and ≤ 0). Thus, there is no loss in generality by considering the compact notation of a system of inequality constraints. It is assumed that all functions may be nonlinear in $\underline{x}$. Convexity or other properties may be assumed when needed. Note that in most real world problems $N \gg n$. This can be written more compactly in vector notation as:

$$\text{MIN } \underline{f}(\underline{x})$$

$$\text{s.t. } \underline{g}(\underline{x}) \leq \underline{0}$$

where $\underline{x} \in R^N$ is the decision vector, $\underline{f}: R^N \rightarrow R^n$ is the objective function vector, $\underline{g}: R^N \rightarrow R^m$ is the constraint vector, and $\underline{0} \in R^m$ is a vector whose elements are all zero. The meaning of minimizing a vector will be discussed in the next section. The definition of " $\leq$ " must also be clarified for vectors:

Definition 1:

For any two vectors, $\underline{y} \in R^k$ and $\underline{z} \in R^k$, $\underline{y} \leq \underline{z}$ if and only if $y_i \leq z_i$ for all $i = 1, 2, \ldots k$, where the subscript i denotes the ith element of the vector.

The constraints $\underline{g}(\underline{x}) \leq \underline{0}$ determine a feasible set T of values for the decision vector $\underline{x}$; $T = \{\underline{x} | \underline{g}(\underline{x}) \leq \underline{0}\}$. Each vector $\underline{x} \in T$ determines a unique value $\underline{f}(\underline{x})$; thus there exists a set S of feasible values for $\underline{f}(\underline{x})$; $S = \{\underline{f}(\underline{x}) | \underline{x} \in T\}$. The multiple objective problem can be considered as:

$$\text{Min } \underline{f}(\underline{x}) \qquad\qquad \text{MIN } \underline{f}(\underline{x})$$
$$\text{or as}$$
$$\text{s.t. } \underline{f}(\underline{x}) \in S \qquad\qquad \text{s.t. } \underline{x} \in T$$

This duality will be useful for understanding the various solution methods presented later. The following examples will serve to illustrate some of these ideas:

EXAMPLE 1:

$$\text{MIN } f_1 = x_1$$

$$\text{MIN } f_2 = 10 - x_1 - x_2$$

$$\text{s.t. } 0 \leq x_1 \leq 5 \text{ (or } g_1 = x_1 - 5 \leq 0, \ g_2 = -x_1 \leq 0)$$

$$0 \leq x_2 \leq 5 \text{ (or } g_3 = x_2 - 5 \leq 0, \, g_4 = -x_2 \leq 0)$$

For this problem T (the feasible set for $\underline{x}$) and S(the feasible set for $\underline{f}(\underline{x})$) are shown in figures 1-1-a and 1-1-b respectively.

1.4   TERMINOLOGY AND CONCEPT OF NON-INFERIOR SOLUTIONS

It is important here to define some of the terms used in multi-objective analysis.   First, the definition of optimal is different than for the case of a single objective function:

Definition 1-2:

An optimal solution is one which attains the minimum value of all of the objectives simultaneously; $\underline{x}^*$ is an optimal solution to the problem MIN $\underline{f}(\underline{x})$ s.t. $\underline{x} \in$ T if and only if $\underline{x}^* \in$ T and $\underline{f}(\underline{x}^*) \leq \underline{f}(\underline{x})$ for all $\underline{x} \in$ T.

Optimal solutions are also known as superior solutions.   In general there is no optimal solution to a multi-objective problem.   In example 1, the minimum value of f_1 is 0 which occurs at $x_1 = 0$, while the minimum value of f_2 is 0 which occurs at $x_1 = x_2 = 5$;   thus these two minima cannot be attained simultaneously.

Let $\overline{f}_i$ be the solution to

$$\min_{\underline{x}} f_i(\underline{x})$$

Subject to

$$g_k(\underline{x}) \leq 0, \, k = 1, 2, \ldots, m$$

$$\text{for all } i = 1, 2, \ldots, n$$

i.e., $\overline{f}_i$ is the global minimum of the ith objective function while ignoring all the other (n-1) objectives. Looking at this in the functional   space, if $\overline{f} = (\overline{f}_1, \overline{f}_2, \ldots \overline{f}_n)$ where $\overline{f}_i$ is as defined above, then $\overline{f} \notin$ S means that no optimal solution exists. For example 1, $\overline{f}=(0,0)$ which is not in S as can be seen from figure 1-1-b.   In the following example there is an optimal solution:

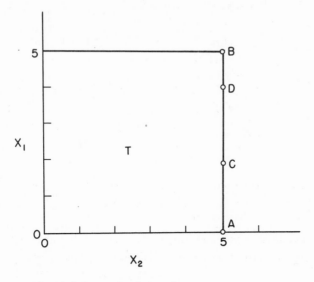

Figure 1-1-a. Decision Space

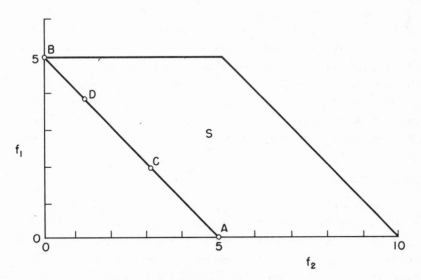

Figure 1-1-b. Functional Space

Figure 1-1. Decision and Functional Spaces for Example 1

EXAMPLE 2:

$$\text{MIN } f_1 = x_1$$

$$f_2 = x_1 + x_2$$

$$g_1 = -x_1 \leq 0$$

$$g_2 = -x_2 \leq 0$$

The minimum value of each objective is zero and these can be attained simultaneously when $x_1 = x_2 = 0$. The optimal solution is $\underline{x}^* = (0,0)$; $\underline{f}(\underline{x}^*) = (0,0)$.

There has been some work done[12] on determining when optimal solutions to a multiple objective problem will exist. Since, however, optimal solutions generally do not exist, one must be satisfied with obtaining non-inferior solutions.

The concept of non-inferior solutions, also known as Pareto optimum or efficient solutions, is basic to economics in general and particularly for competitive equilibrium. Koopmans[13] defined an efficient point for multiobjective functions in economics as follows: "A possible point in the commodity space is called efficient whenever an increase in one of its coordinates (the net output of one good) can be achieved only at the cost of a decrease in some other coordinate (the net output of another good)." Kuhn and Tucker[14] extended the theory of nonlinear programming for one objective function to a vector minimization problem and introduced necessary and sufficient conditions for non-inferior solutions. A formal definition of a non-inferior solution is given below:

Definition 1-3:

A non-inferior solution is one in which no decrease can be obtained in any of the objectives without causing a simultaneous increase in at least one of the other objectives; $\underline{x}^*$ is a non-inferior solution to the problem MIN $\underline{f}(\underline{x})$ s.t. $\underline{x} \in T$ if and only if there does not exist any $\underline{x} \in T$ such that $\underline{f}(\underline{x}) \leq f(\underline{x}^*)$ and $f_i(\underline{x}) < f_i(\underline{x}^*)$ for some $i = 1, 2, \ldots, n$. This solution is obviously not unique. We define the non-inferior set as $NI = \{\underline{f}(\underline{x}) | \underline{x}$ is a non-inferior solution$\}$.

In example 1, the solutions $x_2 = 5$, $0 \leq x_1 \leq 5$ are all non-inferior

solutions. In the functional space (figure 1-1-b) this corresponds to the
line $f_2 = 5 - f_1$. It has been shown[15] that all $\underline{f}(\underline{x})\epsilon NI$ must lie on the
boundary of S.   This is obvious, since for any point in the interior, a
reduction could be achieved in one objective without changing the others
by moving in a negative direction parallel to that axis as far as possible,
that is, until a boundary is reached.   Necessary conditions for a point to
be in the non-inferior set have been developed[16]; the methods to be described subsequently, however, are in general simpler to use than direct
application of these conditions in determining the non-inferior set.

It is easily verified that for any two convex functions, the non-
inferior set is continuous. For example, every point in the interval between the minima $[x_1{}^*, x_2{}^*]$ of the bicriterion problem depicted in Figure
1-2, is a non-inferior point.   For nonconvex functions, however, the non-
inferior set may be non-connected.

Since the non-inferior set is on the boundary of S, it forms a surface in R^n which can be described by $T^*(f_1, f_2,...,f_n) = 0$.   This can be
solved to get $f_i{}^*(f_1,...,f_{i-1}, f_{i+1},...,f_n)$ called the trade-off functions
since they show how much the value of f_i must change to stay in the non-inferior set when the values of the other objectives change.   The rate of
change of the tradeoff function with respect to f_j also forms useful functions which are called the tradeoff rate functions T_{ij}; $T_{ij}(f_1,...,f_{i-1},$
$f_{i+1},...,f_n) = \partial f_i{}^*/\partial f_j$.

Some authors have modified the definition of a non-inferior solution
to exclude those points where a first order improvement in one objective
can be made at the expense of only a second order degradation in another[17];
that is, the points where any of the tradeoff rate functions are either
zero or infinite are not properly non-inferior solutions.

In most cases, the determination of the non-inferior set is not sufficient; the systems analyst must choose one decision which is by some definition "best".   Thus, additional criteria must be introduced to distin-
guish the "best" of the non-inferior solutions.   Although some authors re-
tain the term optimal for this "best" solution, the word "preferred" will
be used in this book to avoid ambiguity.

Definition 1-4:

A preferred solution is a non-inferior solution which is chosen as
the final decision through some additional criteria.

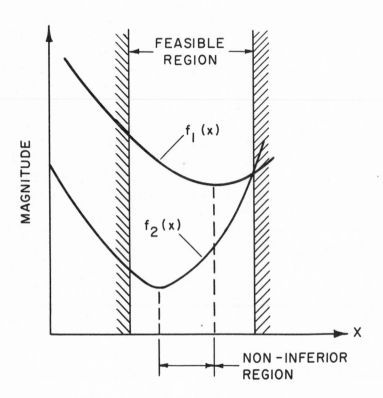

Figure 1-2.   Non-Inferior Points for Convex Bicriterion Problem.

For each value $\underline{f}(\underline{x}) \in R^n$ there is some benefit that would accrue to society from $f_i(\underline{x})$ units of each objective i = 1, 2, ...,n.   This benefit is called the utility function (u: $R^n \rightarrow R^1$).   Since each objective is being minimized, it must be true that it is preferable to have less of each; thus the utility function is monotonically decreasing with respect to each objective. Utility functions are also known as social preference or social welfare functions.

Consider the surfaces of equal utility in $R^n(u(\underline{f})$ = constant). Society is indifferent between any points on these surfaces, and thus they are called social indifference or iso-preference surfaces.   These will be useful later in finding preferred solutions.

The points where the social indifference surfaces are tangent to the non-inferior set are known as the indifference band.

Definition 1-5:

The indifference band is defined to be the subset of the non-inferior set where the improvement of one objective function is equivalent in the mind of the decision maker to the necessary degradation of the others.

1.5   OVERVIEW OF BOOK

The next chapter attempts to give a comprehensive survey of the multiple objective problem and the various approaches and techniques available for the solution of such problems.   Chapter 3 will present still another approach to solving multiple objective problems - the surrogate worth trade-off (SWT) method[18].   Chapter 4 describes some computational efficiencies which can be implemented for two-objective problems in the SWT method. Three algorithms for implementing the SWT approach in static two-objective problems are presented; one employing the ε-constraint approach in solving multiple objective problems; one using Lagrange multipliers in a variation of the parametric approach, and one using a combination of these two.   Chapter 5 modifies the results of chapter 4 in order to apply the SWT to dynamic optimization problems ; included are three analogous algorithms. Chapters 6 and 7 modify. the results of the previous two chapters to encompass problems with more than two objectives (for static and dynamic cases, respectively), and investigate the relationships between the various worth functions W_{ij}, i = 1, 2, ...,n, j = 1, 2, ...,n, i ≠ j. Chapter 8 applies **the** SWT method to **three problems in water resources. Chapter 9 applies the SWT**

method to water quality problems. Chapter 10 discusses the incorporation of sensitivity, irreversibility and risk as multiple objectives in
water resources systems. The final chapter summarizes the major themes of
the book, and indicates areas where its implementation may prove fruitful.

FOOTNOTES

1.   Discussions and examples of these strategies can be found in Kneese
      and Bower [1968], Haimes et al [1972] and others.

2.   The original development can be found in Haimes and Hall [1974].

3.   Multiproject analysis is discussed by many authors including Maass
      et al [1962], Howe and Easter [1971], Hall and Dracup [1970], Isard
      et al [1972] and Haimes and Hall [1974].

4.   The first discussion of a vector of objectives is in Kuhn and Tucker
      [1950].

5.   See Corps of Engineers [1972].

6.   See Miller and Byers [1973].

7.   See Cohon and Marks [1973].

8.   See Major [1969].

9.   See O'Riordan [1973].

10.   See Monarchi et al [1973].

11.   Cohon [1973] provides classifications for the various techniques as
      well as studying their applicability to water resources problems.

12.   Athans and Geering [1973] provide necessary and sufficient condi-
      tions for the existence of superior solutions.

13.   See Koopmans [1951].

14.   Again see Kuhn and Tucker [1950].

15.   A formal proof is given in Reid and Citron [1971].

16.   Different forms can be found in Kuhn and Tucker [1950], Chu [1970]
      and DaCunha and Polak [1967].

17.   The most general definition of proper non-inferiority is given by
      Geoffrion [1968]. Other definitions can be found in Kuhn and Tucker
      [1950] and Klinger [1964].

18.   The original development can be found in Haimes and Hall [1974].

REFERENCES

1.   Athans, M., and Geering, H. P., "Necessary and Sufficient Conditions
         for Differentiable Nonscalar-Valued Functions to Attain Extrema,"
         IEEE Transactions, vol. AC - 18, no. 2, 1973.

2.   Chu, K. C., "On the Non-inferior Set for Systems with Vector-Valued
         Objective Functions," IEEE Transactions, vol. AC-15, no. 5, 1970.

3.   Cohon, J. L., "An Assessment of Multiobjective Solution techniques
         for River Basin Planning Problems," Ph.D. Dissertation, M. I. T.,
         1973.

4.   Cohon, J. L., and Marks, D. H., "Multiobjective Screening Models and
         Water Resource Investment" Water Resources Research, vol. 9, no.
         4, 1973.

5.   Corps of Engineers, "N.A.R. Water Resources Study," Appendix T,1972.

6.   DaCunha, N. O., and Polak, E., "Constrained Minimization under Vec-
         tor Valued Criteria in Finite Dimensional Spaces," Journal   of
         Math. Anal. and Appl., vol. 19, no. 1, 1967.

7.   ●Geoffrion, A. M., "Proper Efficiency and the Theory of Vector Maxi-
         mization," Journal of Math. Anal. and Appl., vol. 22, no. 3,1968.

8.   Haimes, Y. Y., "Hierarchical Modeling for the Planning and Manage-
         ment of a Total Regional Water Resources System," Presented at
         the IFAC Symposium on Control of Water Resources Systems, Israel,
         September 17 - 21, 1973. Also Automatica, Jan. 1975.

9.   Haimes, Y. Y., and Hall, W. A., "Multiobjectives in Water Resources
         Systems Analysis: The Surrogate Worth Tradeoff Method,"
         Water Resources Research, vol. 10, no. 4, 1974.

10.   Haimes, Y. Y., Kaplan, M. A., and Husar, M. A., "A Multilevel Ap-
         proach to Determine Optimal Taxation for the Abatement of Water
         Pollution," Water Resources Research, vol. 8, no. 4, 1972.

11.   Hall, W. A., and J. A. Dracup, Water Resources Systems Engineering,
         McGraw-Hill Book Company, N. Y., 1970.

12.   Howe, C. W., and K. W. Easter, Interbasin Transfers of Water, Econ-
         omics Issues and Impacts, The John Hopkins Press, Baltimore,1971.

13.　　Isard, W. et al, Methods of Regional Analysis; an Introduction to Regional Science, The M.I.T. Press, Cambridge, Massachusetts,1972.

14.　　Klinger, A., "Vector Valued Performance Criteria," IEEE Transactions, vol. AC - 9, no. 1, 1964.

15.　　Kneese, A. V., and B. T. Bower,　　Managing Water Quality: Economics, Technology, Institutions, The Johns Hopkins Press,　Baltimore, Maryland, 1968.

16.　　Koopmans, T. C., "Analysis of Production as an Efficient Combination of Activities," Activity Analysis of Production, Cowles Commission Monograph 13, Edited by T.C. Koopmans, Wiley, N.Y., 1951.

17.　　Kuhn, H. W., and A. W. Tucker, Nonlinear Programming, Proceedings Second Berkeley Symposium on Mathematical Statistics and Probability, pp. 481-492, University of California Press, Berkeley, California, 1950.

18.　　Maass, A. et al, Design of Water-Resource Systems, Harvard University Press, Cambridge, Massachusetts, 1962.

19.　　Major, D.C., "Benefit-Cost Ratios for Projects in Multiple Objective Investment Programs," Water Resources Research, vol. 5, no.6, 1969.

20.　　Miller, W. L., and D. M. Byers, "Development and Display of Multiple Objective Project Impacts," Water Resources Research, vol. 9, no. 1, 1973.

21.　　Monarchi, D. G., C.C. Kisiel, and L. Duckstein. "Interactive Multiobjective Programming in Water Resources," Water Resources Research, vol. 8, Nov. 4, 1973.

22.　　O'Riordan, J., "An Approach to Evaluation in Multiple Objective River Basin Planning," Canada Department of Environment,Vancouver, B. C., 1973.

23.　　Reid, R. W., and S.J. Citron,　"On Non-inferior Performance Index Vectors," Journal of Optimization Theory and Applications, vol.7, no. 1, 1971.

Chapter 2

SOLUTION METHODOLOGIES FOR MULTIPLE OBJECTIVE PROBLEMS

2.1   INTRODUCTION

There are basically two approaches to the solution of problems with
multiple objectives. One can either attempt to find the preferred solution
directly, or first generate the non-inferior set and then find the preferred solution from among these.   A third school of thought is that the systems analyst should be concerned only with developing the non-inferior solutions; the decision maker (DM) can then choose on his own which of these
solutions to implement[1]. It seems logical, however, that the DM will desire some sort of further analysis to find the preferred solution; if this
analysis can be quantified and systemized to reduce the subjectivity, then
a more accurate (according to the criteria introduced) solution will be
found and society will presumably be better off.

2.2   UTILITY FUNCTIONS

The first type of direct approach is the utility function approach;
this assumes that the utility function, $u(\underline{f})$, which can be used to commensurate the various objectives with adequate accuracy, is known. The preferred solution is defined as the one which maximizes society's utility; this
can be found directly by solving: MAX $u(\underline{f}(\underline{x}))$ subject to $\underline{x} \in T$.

Since the utility is monotonically decreasing with respect to each
objective, the preferred solution will be an element of the non-inferior
set[2]. The major drawback to this approach is that, in general, the utility
function cannot be determined.   Much work has been done in decision theory
on how individual and societal utility functions may be approximated[3]. Theoretically, a decision maker reflects the desires of his constituents by
some method of aggregating individual preferences.   Many studies assume
additive utilities $(\underline{u}(\underline{f}) = \sum_{i=1}^{n} u_i(f_i))$; the implications of this assumption
have been analyzed extensively[4]; one implication is that the indifference
between f_i and f_j is independent of the values of the other objectives for
all i and j which is generally unrealistic.   The definition of preferred
solution used in this utility function approach will be used again in some

of the approaches to be described subsequently.

2.3   INDIFFERENCE FUNCTIONS

It is generally conceded[5] that indifference functions are easier to determine than the actual utility function since they can be found by ordinal comparisons. It is easier for a decision maker to determine if he prefers $\underline{f}(\underline{x}_1)$ to $\underline{f}(\underline{x}_2)$ than it is to determine how much additional utility is derived from $\underline{f}(\underline{x}_1)$ (as implied in the utility approach). Thus, recent work has been done using indifference curves to find the preferred solution. Indifference functions can be used to make the objectives commensurate since they relate how much increments in one objective are worth in terms of another. Briskin[6] assumed the form of the indifference function was known (exponential) for a two objective (minimize time and cost) problem. The constants in the equation are determined by questioning the DM to find several points on the curve; then substituting for f_2 in terms of f_1 from the indifference equation the preferred solution can be found by solving:

$$\text{Min } f_1(\underline{x}) + f_2(f_1(\underline{x}))$$

$$\text{s.t. } \underline{x} \; \epsilon \; T$$

The problem with this approach is that it assumes the indifference equation will be the same everywhere in the feasible space. This is generally not the case. Another approach using indifference functions is to find where one of the indifference surfaces is tangent to the tradeoff function; this will give the preferred solution in the maximum utility sense.

2.4   LEXICOGRAPHIC APPROACH

The lexicographic approach[7] requires that the objectives be ranked in order of importance by the DM. A preferred solution for this approach is defined to be one which simultaneously minimizes as many of the objectives as possible, starting with the most important and going down the hierarchy. Let $f_1(\underline{x})$ be the most important objective; then the problem MIN $f_1(\underline{x})$ s.t. $\underline{x} \; \epsilon \; T$ is solved for all possible solutions; we call the set of all solutions to this problem y_1. Then the next most important objective $f_2(\underline{x})$ is minimized subject to $\underline{x} \; \epsilon \; y_1$ to find the solution set y_2. This

process is repeated until all n objectives have been considered. If the
solution set y_i at the ith iteration has only one element, then this will
be the solution to the entire problem; the objectives ranked less important
than f_i are ignored by this method.   The rationale for this approach is
that individuals tend to make decisions in this manner[8]. This lexicographic definition of preferred was used by McGrew and Haimes for the problem of
joint system identification and optimization[9]. Note that the solution will
be very sensitive to the ranking by the DM, and thus the analyst should exercise caution in applying this method when two objectives are of nearly
equal importance.

        To illustrate this approach, consider again example 1 from Chapter
1, and assume that the DM has decided that f_1 is the most important objective. First, minimize $f_1 = x_1$ s.t. $0 \leq x_1 \leq 5$, $0 \leq x_2 \leq 5$ and get the sol-
ution set $y_1 = \{(x_1, x_2) | x_1 = 0, 0 \leq x_2 \leq 5\}$. Then, minimize $f_2 = 10 - x_1$
$- x_2$, s.t. $x_1 = 0$, $0 \leq x_2 \leq 5$.   The solution is $x_1 = 0$, $x_2 = 5$, $f_1 = 0$, f_2
$= 5$; this is the preferred solution and is shown as point A in figures 1-1-
a and 1-1-b.

        A variation of this method was proposed by Waltz[10]; after the prim-
ary objective is minimized, the second objective is minimized subject to
keeping the first objective within a certain percentage of its optimum
(this percentage is determined by the DM).   The third objective is then
minimized keeping the first two within a certain percentage of the values
found in the previous step. This process is repeated until all the objectives have been considered. This approach reduces the sensitivity somewhat,
but the same caveat is necessary.

2.5   PARAMETRIC APPROACH

        Assume that the relative importance of the n objectives is known and
constant. From a utility viewpoint this implies additive and linear utilities. Then the preferred solution is found by solving:

$$MIN \sum_{i=1}^{n} \Theta_i f_i(\underline{x})$$

$$s.t. \ \underline{x} \in T$$

where $\Theta_i > 0$ are the weighting coefficients which determine the relative

importance of the objectives (usually the Θ_i are normalized so that

$\sum_{i=1}^{n}\theta_i=1$). For simplicity in notation, let $\underline{\theta}^T$ represent the row vector (θ_1, θ_2, ..., θ_n), where the superscript T denotes the transpose operation.

One drawback of this approach, however, is that the proper relative weight to be given to any objective on this scale is usually a function not only of the quantity of that objective produced, but also of the quantities of all other objectives to be produced. Even monetary value is subject to this criticism: To most individuals a second $1,000 is not equally as important or valuable as the first, and the values of each usually depend heavily on the individual's attainment levels of other physical and social objectives.

At least in theory, this means that the method should require the consensus of the experts on the "price" of each objective for all possible combinations of levels of objectives attained. This complexity can be eliminated if the decision only adds or subtracts a negligibly small increment to the total of any objective. It could then be argued that, so long as all the experts were fully aware of the status quo, the error due to considering the values of each investment of each objective as independent quo should not be serious. When dealing with national policy issues, however, either the proposed or expected changes in objective levels are trivial (in which case the analysis is unimportant) or the decisions are intended to produce major increments in the various objectives. In the latter case the problem of commensuration reappears with great significance.

The parametric approach was used by Major[11] in modifying cost-benefit analysis for water resources planning. Although the assumption of known and constant values for the relative importance of the objectives is generally not acceptable[12], this method can be used to generate points in the non-inferior set by utilizing various values of $\underline{\theta}$; this was first suggested by Everett[13] for resource allocation problems. For convex problems (that is, when the tradeoff function is convex), the parametric approach generates the entire non-inferior set[14]. Reid and Vemuri[15] found the tradeoff function f_1^* as a function of $\underline{\theta}$ when the objective functions are posynomials, but Reid and Citron[16] found that even simple systems generally give incredibly complex functions for $f_1^*(\underline{\theta})$. Geoffrion[17] developed a parametric procedure for finding the entire non-inferior set for the case of maximizing two concave functions with concave constraints. The parametric approach has been used in linear problems in developing a modified simplex method to determine the entire non-inferior set[18]. Geoffrion[19] used the parametric method to find the non-inferior decision vector $\underline{x}^*$ as a function of α for the two objective case where $\alpha = \theta_1$ and $1-\alpha=\theta_2$. The utility function (assumed known) could then be found as a function of α and then maximized to find the preferred solution. McGrew and Haimes[20] used the parametric approach iteratively to converge to the lexicographic preferred solution for the joint identification and optimization problems.

To illustrate the direct form of the parametric approach, consider again example 1, and suppose the DM has decided that objective f_2 is three times as important as objective f_1. The problem then becomes:

$$MIN\ x_1 + 3(10\ - x_1 - x_2) = 30 - 2x_1 - 3x_2$$

$$s.t.\ 0 \leq x_1 \leq 5,\ 0 \leq x_2 \leq 5$$

The solution to this problem is $x_1 = 5$, $x_2 = 5$, $f_1 = 5$, $f_2 = 0$; this preferred solution is shown as point B in figures 1-1-a and 1-1-b.

The parametric approach can be interpreted geometrically as follows. The set $L = \{\underline{f}(\underline{x}) | \underline{\theta}^T \cdot \underline{f}(\underline{x}) = c\}$ (where c is a constant), defines a hyperplane[21] in R^n with outward normal - $\underline{\theta}$. The minimization of $\underline{\theta}^T \cdot \underline{f}(\underline{x})$ can be viewed as moving this hyperplane L with fixed $\underline{\theta}$ in a negative direction as far as possible keeping $L \cap S$ non-null. This minimum will generally occur where L is tangent to S (but not always; this will cause problems later). This is depicted for the two objective case in figure 2-1; L is a line with slope - θ_2/θ_1; the minimum for this value of $\underline{\theta}$ occurs at point A.

For non-convex problems, when one tries to find the entire non-inferior set, the problem of duality gaps arises; some points in the non-inferior set cannot be found for any value of $\underline{\theta}$. Consider the point A in figure 2 - 2. The line L which is tangent at A with slope - θ_2/θ_1 can be moved farther in a negative direction until it is tangent at point B. Thus the parametric approach with this value of $\underline{\theta}$ will find points B and C, but not point A.

Geometrically speaking, points in the non-inferior set which do not have a supporting hyperplane cannot be found by the parametric method. The term duality gap comes from the fact that the parametric problem MIN $\underline{\theta}^T \cdot f(\underline{x})$ is related to the dual of the problem MIN $f_1(\underline{x})$ s.t. $f_j(\underline{x}) \leq y_j$ $j = 2$, $3, \ldots, n$ where the y_j are constants. These gaps can be explained in terms of duality theory[22].

2.6   THE ε-CONSTRAINT APPROACH

The direct form of the ε-constraint approach[23] requires the DM to

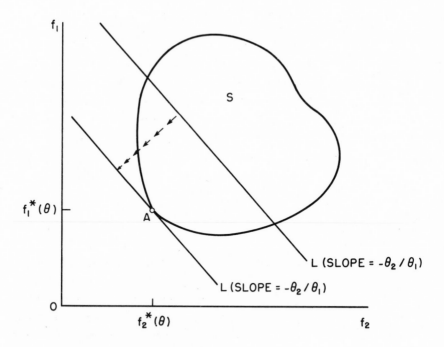

Figure 2-1   Parametric Approach

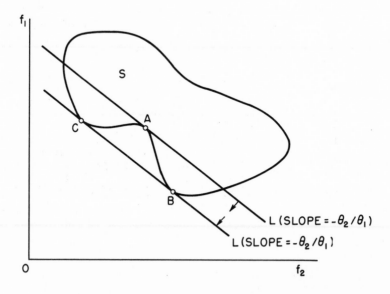

Figure 2-2   Duality Gaps in the Parametric Approach

specify the maximum allowable levels (ε_2, ε_3, ... ε_n) for the n - 1 objectives (f_2, f_3, ...,f_n); the preferred solution is the one which solves the
following problem:

$$\text{Min } f_1(\underline{x})$$

$$\text{s.t. } f_j(\underline{x}) \leq \varepsilon_j \; j = 2, 3, ...,n$$

$$\underline{x} \; \varepsilon \; T$$

Of course, any objective could be chosen as f_1. From a utility viewpoint,
this approach says that the benefit to society from objective f_j is constant as long as the level ε_j is not exceeded, but becomes infinitely harm-
ful above this level. In other words the utility function is additive with

$$u_j(f_j) = \begin{cases} \text{constant; } f_j \leq \varepsilon_j \\ \\ -\infty \; ; \; f_j > \varepsilon_j \end{cases}$$

This approach can also be interpreted in terms of the lexicographic approach if ε_j is interpreted as the satisfaction level of the jth objective,
and f_1 as the least important objective. Then the set y_{n-1} of solutions
to the n-1st stage would be $\{\underline{x} | f_j(\underline{x}) \leq \varepsilon_j, j = 2,3, ..., n,$ and $\underline{x} \; \varepsilon \; T;\}$
thus the ε-constraint approach can be interpreted as the nth iteration in
the lexicographic approach.

The determination of the maximum levels as well as the assumption
about this form of preference are often questionable in real problems; however, this method can be used to generate the non-inferior set for all
types of problems by varying the values of the ε_j,j = 2, 3, ...,n. Specific methods for achieving this will be described in later chapters. This
approach has been used for the joint identification and optimization problem,[24] and for water resources problems with linear objectives and constraints[25]. Haimes et al[26] prove that this approach does give non-inferior
solutions for the two objective cases. Pasternak and Passy[27] used a combination of the parametric and ε-constraint approaches to find the preferred (maximum utility) solution to a non-convex, two objective, 0-1 integer
programming problem. Due to the non-convexity, the parametric approach
could not be used alone, and this mixed approach was found to be more efficient than using a straight ε-constraint approach.

To illustrate the direct form of the ε-constraint approach, consider again example 1, and suppose that the DM has decided that the maximum level society can tolerate of f_2 is 3 units. The problem is then:

$$\text{Min} \quad x_1$$

$$\text{s.t.} \quad 10 - x_1 - x_2 \leq 3$$

$$0 \leq x_1 \leq 5$$

$$0 \leq x_2 \leq 5$$

The solution to this problem is $x_1 = 2$, $x_2 = 5$, $f_1 = 2$, $f_2 = 3$. This preferred solution is shown as point C in figures 1-1-a and 1-1-b.

Geometrically, this approach adds additional constraints which reduce the feasible decision space T, or equivalently the feasible functional space S. Define $T_j' = \{\underline{x} | f_j(\underline{x}) \leq \varepsilon_j\}$ for $j = 2, 3, \ldots, n$; then $S_j' = \{\underline{f}(\underline{x}) | \underline{x} \varepsilon T_j\}$ for $j = 2, 3, \ldots, n$ and $S' = S \cap S_2' \cap S_3' \cap \ldots \cap S_n'$. The problem is now

$$\text{MIN} \quad f_1(\underline{x})$$

$$\text{s.t.} \quad \underline{f}(\underline{x}) \varepsilon S'$$

Each constraint $f_j(\underline{x}) \leq \varepsilon_j$ defines the half-space in R^n on the negative side of a hyperplane perpendicular to the f_j axis at $f_j = \varepsilon_j$. The intersection of all of these half-spaces with S gives the new feasible space S'. This is depicted for the two objective case in figure 2-3: $S' = S \cap S_2$ where s' is the half-plane to the left of the line $f_2 = \varepsilon_2$. Note that this approach can determine the entire non-inferior set even for non-convex problems.

2.7   GOAL PROGRAMMING

The goal programming method requires the DM to set goals that he would like each objective to attain. A preferred solution is then defined as the one which minimizes the deviations from the set goals. Denote the vector of goals set by the DM for the objectives by $\hat{\underline{f}}$; then the mathematical formulation of the problem is

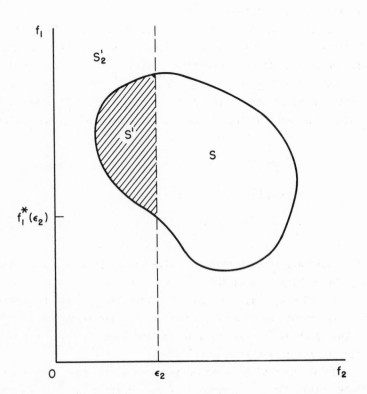

Figure 2-3.   ε-Constraint Approach.

$$\text{MIN } ||\underline{f}(\underline{x}) - \hat{\underline{f}}||$$

$$\text{s.t. } \underline{x} \in T$$

where $||\cdot||$ denotes any norm[28]. Note that the goal vector $\hat{\underline{f}}$ does not have to be in the feasible set S; for example, in example 1 the goal vector could be $\hat{\underline{f}}^T = (0, 0)$. In fact, if the goal vector is in S, then this method may yield an inferior solution.

This approach was developed by Charnes and Cooper[29] for linear problems. Using the sum of the absolute values of the deviations as the norm, they keep the problem linear by defining vectors of slack variables $\underline{y}^+ \geq \underline{0}$ and $\underline{y}^- \geq \underline{0}$ such that $\underline{f}(\underline{x}) - \hat{\underline{f}} = \underline{y}^+ - \underline{y}^-$; y_i^+ (the ith component of $\underline{y}^+$) is then the over-attainment of the ith objective, and y_i^- (the ith component of $\underline{y}^-$) is the under-attainment of the ith objective. The problem then becomes:

$$\text{MIN } \sum_{i=1}^{n} y_i^+ + y_i^-$$

$$\text{s.t. } \underline{f}(\underline{x}) - \hat{\underline{f}} = \underline{y}^+ - \underline{y}^-$$

$$\underline{x} \in T$$

To illustrate this approach, consider again example 1 and assume $\hat{\underline{f}}^T = (0, 0)$. The problem is then MIN $y_1^+ + y_2^+ + y_2^-$ s.t. $x_1 = y_1^+$, $10 - x_1 - x_2 = y_2^+ - y_2^-$, $0 \leq x_1 \leq 5$, $0 \leq x_2 \leq 5$. Note that y_1^- was not necessary since $f_1 = x_1$ cannot go below 0 (f_1 cannot be under-attained). The solution to this problem is that any point on the line $f_1 + f_2 = 5$ minimizes this objective and is a preferred solution.

A special case of this method is known as the mean square approach. This assumes that the ith component of the goal vector $\hat{\underline{f}}$ will be $\overline{f}_i$, the minimum value of $f_i(\underline{x})$ s.t. $\underline{x} \in T$, and uses a least square norm. Salukvadze[30] applied this method to optimal control problems; it can be shown[31] that the solution to this problem is automatically in the non-inferior set. This approach eliminates the decision maker entirely and thus this definition of preferred solution will generally not maximize the benefit to society.

In general, the problem with goal programming is that an equal importance is placed on each objective. If weighting factors are introduced to counteract this, then the problem of determining the weights for any

real problem arises.   Just as in the parametric case, the non-inferior set
can be generated by varying the weighting factors, but it can be shown[32]
that this   method suffers from the same duality gap problem as the parametric approach when the problems are non-convex.

2.8   THE GOAL ATTAINMENT METHOD

A variant of the goal   programming method is the goal   attainment
method[33].   In this approach, a vector of weights $\underline{\omega}$ relating the relative
under or over attainment of the desired goals must be determined by the DM
in addition to the goal vector $\hat{\underline{f}}$.   The preferred solution solves the problem:

$$\text{MIN } z$$

$$\text{s.t. } \underline{f}(\underline{x}) - \underline{\omega} \, z \le \hat{\underline{f}}$$

$$\underline{x} \in T \quad ; \quad \underline{\omega} > \underline{0}$$

where z is a scalar variable unrestricted in sign.   This approach can also
be used to generate the non-inferior set; using $\hat{\underline{f}}$ as in the mean square ap-
proach   (the ith component of $\hat{\underline{f}}$ is $\overline{f}_i$,   the minimum value of $f_i(\underline{x})$ s.t. $\underline{x} \in$
T), the entire non-inferior set can be found by varying $\underline{\omega}$, even for non-
convex problems. Again $\underline{\omega}$ is generally normalized so that $\sum\limits_{i=1}^{n} \omega_i = 1$.   This
approach has been successfully applied to static problems (economic dispatch
in power   system control) and dynamic problems (the load-frequency control problem in regulator design).

This approach is depicted for the two objective case in figure 2-4 .
$\underline{\omega}$ and $\hat{\underline{f}}$ fix the direction of the vector $\hat{\underline{f}} + \underline{\omega} z$, and the minimum value of
z occurs where this vector first intersects S.

To illustrate the use of this method consider again example 1. Assume that the goal vector set by the DM is $(0, 0)^T$ and that he decides the
relative over-attainment of f_2 should be 1/3 the over-attainment of f_1 .
Then $\underline{\omega}^T = (.75, .25)$ and the goal attainment problem becomes:

$$\text{Min } z$$

$$\text{s.t. } x_1 - .75 \, z \le 0$$

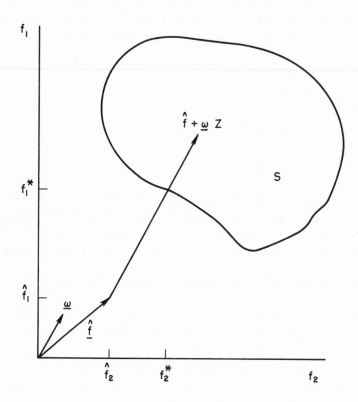

Figure 2-4.   Goal Attainment Method.

$$10 - x_1 - x_2 - .25 z \leq 0$$

$$0 \leq x_1 \leq 5$$

$$0 \leq x_2 \leq 5$$

The solution to this problem is $z = 5$, $x_2 = 5$, $x_1 = 15/4$, $f_1 = 15/4$, $f_2 = 5/4$. This is depicted as point D in figures 1-1-a and 1-1-b.

2.9   ADAPTIVE SEARCH APPROACH

This approach[34] is useful if one is only interested in determining non-inferior values. One starts with an initial non-inferior vector $\underline{x}_0$ in the decision space; the ith component of $\underline{x}_0$ is found by solving:

$$\text{MIN } f_i(\underline{x})$$

$$\text{s.t. } \underline{x} \in T$$

Then new solutions are generated with the following recursive formula:

$$\underline{x}_{i+1} = \underline{x}_i - a_i \underline{J}^T(\underline{x}_i) \underline{\omega}_i + \underline{c}_i$$

where a_i controls the step size, $\underline{J}$ is the Jacobian matrix of partial derivatives of the objectives with respect to the decision variables, $\underline{\omega}_i$ controls the direction and $\underline{c}_i$ controls the feasibility of the solution. Each new solution is then checked for non-inferiority as follows; if any two of the gradients are of opposite sign, or the point is on the boundary of T, then it is a candidate for non-inferiority. After a large number of steps (e.g. 1000) are completed, the candidates for non-inferiority are compared; those that are too close together or inferior are eliminated; the remaining ones should be a good approximation to the non-inferior set. At this point, a regression or interpolation can be used to determine an analytic equation; it is generally easier to accomplish this in the functional space. The drawbacks of this approach are that the computational effort can become immense when there are many decisions, and that no means of choosing a preferred solution from the non-inferior ones is presented.

2.10   INTERACTIVE APPROACHES

Much work has been done recently in interactive approaches to finding preferred solutions; in these methods a search procedure is used to
find the preferred solution with questions being asked of the DM at each
step of the search in order to determine a new estimate of the solution.
An interactive approach developed by Geoffrion and examined by Feinberg
uses indifference functions to determine the maximum utility preferred solution[35]. Since the social indifference function is a surface of constant
utility, the normal to this surface is the direction of greatest increase
of the utility function.   Thus one can pick some initial point $\underline{x} \in T$, and
by questioning the DM, the social indifference function around this point
can be found; the normal to this function is the direction to move in pick-
ing the next approximation for $\underline{x}$.   The step size is found by calculating
$\underline{f}(\underline{x})$ for different step sizes (keeping $\underline{x} \in T$) and by questioning the DM to
find the one he prefers most.   Termination occurs when the improvement be-
tween steps is less than some specified value. The attributes of this approach are that no assumptions about the form of the utility function are
necessary, and that the DM need only consider relative preferences[36].

2.11   OTHER APPROACHES

There are a number of other approaches to multiple objective problems which can only be mentioned here due to space limitations.   One of
those is to model the decision-making process itself[37], using techniques
such as game theory to determine the preferred solution; these models, however, tend to be computationally prohibitive. Other authors[38] have considered the multi-objective problem for the case where there are only two or
three non-inferior points from which the preferred solution must be chosen,
developing specialized techniques for these situations. Another approach[39]
is to generate a stronger partial ordering of the non-inferior solutions;
this method allows for uncertainties on the part of the DM, but as a result
will only eliminate some of the non-inferior points. Another type of problem is decision making under uncertainty.   Much work has been done in the
area[40], but this goes beyond the scope of this text.

FOOTNOTES

1. This was suggested by Zadeh [1963].
2. This was shown by Geoffrion [1967]; also see definition 4.
3. Many studies in economic theory have been devoted to this question; see, for example, Arrow [1963] or Bergson [1954].
4. See Fishburn [1967], or Keeney [1972] among others.
5. A good discussion of this question can be found in Arrow [1963].
6. See Briskin [1966].
7. This was first introduced by Georgescu-Roegen [1954].
8. There is a relationship between the lexicographic approach and utility theory; see Robinson and Day [1972].
9. See McGrew and Haimes [1974].
10. See Waltz [1967].
11. Major [1969] was the first to use multiple objective analysis for problems in water resources planning.
12. See the criticism of Freeman and Haveman [1970] for example.
13. See Everett [1963].
14. This is proven by Geoffrion [1968].
15. The problem to which Reid and Vemuri [1971] apply their results is discussed further in Chapter 8.
16. See Reid and Citron [1971].
17. See Geoffrion [1966].
18. Various authors have proposed ways of accomplishing this. See Yu and Zeleny [1973], Sengupta [1972], and Evans and Steuer [1972].
19. See Geoffrion [1967].
20. See McGrew and Haimes [1974].
21. A hyperplane is the generalization of a plane into n dimensions; a one dimensional hyperplane is a line.
22. For an explanation of duality gaps see Lasdon [1968] or Gembicki [1973].
23. A good description of this approach can be found in Haimes [1973b].
24. Applications to the joint identification and optimization problems can be found in Haimes and Wismer [1972] and Olagundoye and Haimes [1973].
25. This technique was used by Cohon and Marks [1973] and Miller and Byers [1973].
26. See Haimes et al [1971].

27.   See Pasternak and Passy [1972].

28.   The general use of norms is discussed by Salukvadze [1974] and Yu and Lietmann [1974].

29.   The original development of goal programming is given in Charnes and Cooper [1961].

30.   See Salukvadze [1971].

31.   This is proven by Huang [1972].

32.   The proof can be found in Gembicki [1973].

33.   The original development of the goal attainment method is given by Gembicki [1973].

34.   The original development of the adaptive search approach is given in Beeson and Meisel [1971].

35.   This interactive approach is described in detail by Geoffrion [1970] and Feinberg [1972].

36.   Descriptions of other interactive approaches can be found in Roy [1971], Feinberg [1972], and Cohon [1973].

37.   A number of these models are described in Cohon [1973] and Cochrane and Zeleny [1973].

38.   A good example is by Maier-Rothe and Stankard [1970];   some others are described in MacCrimmon [1972].

39.   This approach is described in Roy [1971] and Cohon [1973].

40.   The general problem is described in Raiffa [1968]; other models can be found in Cochrane and Zeleny [1973].

REFERENCES

1.   Arrow, K. J., Social Choice and Individual Values, John Wiley and Sons, N.Y., 1963.

2.   Beeson, R.M. and Meisel, W.S., "The Optimization of Complex Systems with Respect to Multiple Criteria," Proceedings: Systems, Man and Cybernetics Conference, Anaheim, Cal., 1971.

3.   Bergson, A., "On the Concept of Social Welfare," Quarterly Journal of Economics, vol. 68, 1954.

4.   Briskin, L.E., "A Method of Unifying Multiple Objective Functions," Management Science, vol. 12B, no. 10, 1966.

5.   Charnes, A., and Cooper, W.W., Management Models and Industrial Application of Linear Programming, vol. 1, John Wiley and Sons, N.Y., 1961.

6.   Cochrane, J. L., and Zeleny, M., Multiple Criteria Decision Making,

U. of South Carolina Press, Columbia, S.C., 1973.

7.   Cohon, J. L., and D. H. Marks, "Multiobjective Screening Models
      and Water Resource Investment," Water Resources Research, vol. 9,
      no. 4, 1973.

8.   Cohon, J. L., "An Assessment of Multiobjective Solution Techniques
      for River Basin Planning Problems," Ph.D. Dissertation, M. I. T.
      1973.

9.   Evans, J.R. and Steuer, R.E., "Generating Efficient Extreme Points
      in Linear Multiple Objective Programming: Two Algorithms and
      Computing Experience," Presented at Seminar on Multiple Criteria
      Decision Making, U. of South Carolina, 1972.

10.   Everett, H., III, "Generalized Lagrange Multiplier Method for Sol-
      ving Problems of Optimum Allocation of Resources," Operations
      Research, vol. 11, 1963.

11.   Feinberg, A., "An Experimental Investigation of an Interactive Ap-
      proach for Multi-criterion Optimization with An Application to
      Academic Resource Allocation," Western Management Science Insti-
      tute, Working paper no. 186, 1972.

12.   Fishburn, P. C., "Methods of Estimating Additive Utilities," Manage-
      ment Science, vol. 13, no. 7, 1967.

13.   Freeman, A. M., III, and Haveman, R. H., "Benefit-Cost Analysis and
      Multiple Objectives: Current Issues in Water Resources Planning,"
      Water Resource Research, vol. 6, no. 6, 1970.

14.   Gembicki, F., "Vector Optimization for Control with Performance and
      Parameter Sensitivity Indices," Ph.D. Dissertation, Case Western
      Reserve University, 1973.

15.   Geoffrion, A. M., "Strictly Concave Parametric Programming," Manage-
      ment Science, vol. 13, no. 3, 1966.

16.   Geoffrion, A.M. "Solving Bicriterion Mathematical Programs," Opera-
      tions Research, vol. 15, no. 1, 1967.

17.   Geoffrion, A.M., "Proper Efficiency and the Theory of Vector Maxi-
      mization," Journal of Math. Anal. and Appl., vol. 22, no. 3,
      1968.

18.   Geoffrion, A.M., "Vector Maximal Decomposition Programming," Western
      Management Science Institute, Working paper no. 164, 1970.

19.   Georgescu-Roegen, N., "Choice, Expectations, and Measurability,"
      Quarterly Journal of Economics, vol. 64, 1954.

20.   Haimes, Y. Y., "Integrated System Identification and Optimization,"

in Control and Dynamic Systems: Advances in Theory and Applications, vol. 10, Academic Press, Inc., 1973 .

21.   Haimes, Y.Y., Lasdon, L.S. and Wismer, D.A., "On a Bicriterion Formulation of the Problems of Integrated System Identification and System Optimization," IEEE Transactions, vol. SMC-1, 1971.

22.   Haimes, Y. Y., and Wismer, D. A., "A Computational Approach to the Combined Problem of Optimization and Parameter Identification," Automatica, vol. 8, 1973.

23.   Huang, S.C., "Note on the Mean-Square Strategy for Vector Valued Objective Functions," Journal of Optimization Theory and Applications, vol. 9, no. 5, 1972.

24.   Keeney, R. L., "Multiplicative Utility Functions," MI.T. Operation Research Center, Technical Report no. 70, 1972.

25.   Lasdon,L.S. "Duality and Decomposition in Mathematical Programming" IEEE Transactions, vol. SSC-4, no. 2, 1968.

26.   MacCrimmon,K.R., "An Overview of Multiple Objective Decision Making" Presented at Seminar on Multiple Criteria Decision Making, U. of South Carolina, 1972.

27.   Maier-Rothe, C. and Stankard, M.F. Jr., "A Linear Programming Approach to Choosing Between Multi-objective Alternatives," presented at the 7th Mathematical Programming Symposium, the Hague, 1970.

28.   Major, D.C. "Benefit-Cost Ratios for Projects in Multiple Objective Investment Programs," Water Resources Research, vol. 5, no. 6, 1969.

29.   McGrew, D.R., and Haimes, Y. Y., "A Parametric Solution to the Joint System Identification and Optimization Problem," Journal of Optimization Theory and Applications, vol. 13, no. 5, 1974.

30.   Miller, W. L., and Byers, D.M., "Development and Display of Multiple Objective Project Impacts," Water Resources Research, vol. 9, no. 1, 1973.

31.   Olagundoye, O., and Haimes, Y. Y., "The Epsilon-Constraint Approach for Solving Bicriterion Programs," SRC Technical Report. CWRU, 1973.

32.   Pasternak, H., and Passy, Y., "Annual Activity Planning with Bicriterion Functions," Technion mimeograph series no. 110, 1972.

33.   Raiffa, H., Decision Analysis, Addison-Wesley, Reading, Mass, 1968.

34.   Reid, R. W., and Citron, S. J., "On Non-Inferior Performance Index Vectors," Journal of Optimization Theory and Applications, vol.7, no. 1, 1971.

35. Reid, R. W., and Vemuri, V., "On the Non-Inferior Index Approach to Large Scale Multi-Criteria Systems," Journal of the Franklin Institute, vol. 291, no. 4, 1971.

36. Robinson, S.M., and Day, R.H., "Economic Decisions with L^{**} Utility" Social Systems Research Institute, U. of Wisconsin, Paper no.7227 1972.

37. Roy, B., "Problems and Methods with Multiple Objective Functions," Mathematical Programming, vol. 1, no. 2, 1971.

38. Salukvadze, M.E. "Optimization of Vector Functionals: The Programming of Optimal Trajectories," Tbilisi, translated from Avtomatika i Telemekhanika, no. 8, 1971.

39. Salukvadze, M.E. "On the Existence of Solutions in Problems of Optimization under Vector Valued Criteria," Journal of Optimization Theory and Application, vol. 13, no. 2, 1974.

40. Sengupta, S. S., "Probabilities of Optima in Multi-Objective Linear Programmes," presented at Seminar on Multiple Criteria Decision Making at U. of South Carolina, 1972.

41. Waltz, F. M., "An Engineering Approach: Hierarchical Optimization Criteria," IEEE Transactions, vol. AC-12, 1967.

42. Yu, P.L. and Leitmann, G., "Compromise Solutions, Domination Structures, and Salukvadze's Solution," Journal of Optimization Theory and Applications, vol. 13, no. 3, 1974.

43. Yu, P. L., and Zeleny, M., "The Set of all Non-Dominated Solutions in the Linear Cases and a Multi-Criteria Simplex Method," U. of Rochester, Center for System Science, 1973.

44. Zadeh, L. A., "Optimality and Non-Scalar Valued Performance Criteria" IEEE Transactions, vol. AC-8, no. 1, 1963.

Chapter 3

<h1 style="text-align:center">THE SURROGATE WORTH TRADE-OFF METHOD</h1>

The Surrogate Worth Trade-off Method recognizes that optimization theory is usually much more concerned with the relative value of additional increments of the various non-commensurable objectives, at a given value of each objective function, then it is with their absolute values.[1] Furthermore, given any current set of objective levels attained, it is much easier for the decision makers to assess the relative value of the trade-off of marginal increases and decreases between any two objectives than it is for them to assess their absolute values. In addition, the optimization procedure can be developed so that it requires no more than an assessment of whether an additional quantity of one objective is worth more or less than that which may be lost from another, given the levels of each. The ordinal approach can then be utilized with much less concern for the potential distortions in relative evaluations introduced by attempting to commensurate the total value of all objectives concerned.

Since the dimension of the decision space N for most real world problems is generally higher than that of the functional space n (N decisions and n objectives N >>n), a further simplification is to establish decisions in the functional space and then later transform this information into the decision space.

3.1   GENERAL APPROACH

The basic concept of the Surrogate Worth Trade-off method for non-commensurate multi-objective optimization will be explained through a simplified example of commensurate multi-objective optimization. Consider an unconstrained, two objective, one decision variable optimization problem in which both objectives are measurable in the same units, e.g., monetary value.

$$\min f_1(x) + f_2(x)$$

Applying the classical calculus optimization

$$\frac{df_1}{dx} + \frac{df_2}{dx} = 0$$

Thus $df_1/df_2 = -1$ defines optimality subject to the usual necessary and sufficient conditions and tests. It will be noted that df_1/df_2 is the trade-off ratio between objective one and objective two, hence the trade-off ratio at optimality must equal minus one when f_1 and f_2 are in fully commensurate units. Note that it is this ratio of the value of the small increment in f_1 to the value of the resulting increment in f_2 that is significant. Except to the extent that these incremental values depend upon the attained level of both objectives, absolute values of f_1 and f_2 do not appear in the optimality equation. The concern is for the relative value of the increments, given an attained level of achievement of both objectives (whether or not they have the same units of measurement).

Next let f_1 and f_2 be measured in different units or dimensions, e.g., firm water and firm energy from a river-reservoir operations. In this case df_1/df_2 defines the trade-off ratio T_{12}. At optimality, the commensurated value of T_{12} must equal minus one. Since f_1 and f_2 are in different units let T_{12} be multiplied by W'_{12}, the ratio of the true (but unknown) per unit worth of any increment Δf_1 to the true (but unknown) per unit worth of any increment Δf_2 at the known attained levels of satisfaction of objectives f_1 and f_2. If it could be determined, this ratio W'_{12} would be the worth coefficient for the trade-off T_{12}.

By definition, in a non-commensurate problem W'_{ij} cannot be determined for all values of either $\underline{x}$ (decision space) or f_1 and f_2 (objective space); otherwise the objectives could be commensurated and standard optimization techniques applied.

However, consider a "surrogate worth function" W_{12}, which possesses the following properties. First, it has a positive value if the decision maker considers that the true worth of Δf_1 is greater than the true worth of Δf_2. Second, it has a negative value if the opposite is true. In combination these two properties assign the value of zero to any decision which results in indifference; that is, the decision maker, with the available information, cannot determine whether the incremental gain in one objective is or is not preferable to the necessary loss in another. Finally the third property of W_{12} is that it is monotonically consistent in an ordinal sense. That is, a value of +5 represents a stronger feeling that the true worth of Δf_1 is greater than Δf_2 than does a value of +3.

The surrogate worth function now has all the properties needed for its construction and implementation in finding the preferred solution. Using ordinary slope intercept or curve fitting procedures for successive approximation, the zero of this particular surrogate worth function can be quickly found.   By definition of the zero value, such a solution is equivalent to marginal loss equal to marginal gain, hence the following definition of a preferred solution.

Definition 3-1: A preferred solution is defined to be any non-inferior feasible solution which belongs to the indifference band.

To summarize the concept behind the SWT method, a surrogate worth function is substituted for the true (but unknown) worth function which (if known) would commensurate the numerator and denominator of the physical tradeoff ratios T_{ij} .   The surrogate has the property of monotonicity and a value of zero (or other arbitrary number) whenever the value of the numerator of the trade off ratios equals the value of the denominator. Thus when the surrogate worth function has a value of zero, the corresponding solution is within the band of indifference and no other solution can be judged superior to it.

In practice all that is required of the decision maker is to determine whether or not an incremental gain is worth the corresponding incremental loss for any tradeoff T_{ij} , and if not, which is greater. Computational efficiency is gained if he also estimates how far from equal (or indistinguishable) the worth of the proposed results are. Worth need be evaluated only in the relative sense, e.g., whether Δf_i > Δf_j in worth and the improvement between successive trials is "large" or "small".

The computational procedure may be executed in decision space, in objective space or in trade-off ratio space. Since objectives are normally far fewer in number than decision variables, it will usually be preferable to work in objective or trade-off ratio space. Usually the number of objectives to be considered simultaneously are of the order of 10 .or less, while the number of decision variables may be of the order of a thousand. This results in a decided computational advantage for the use of objective space. It is also more realistic for interactions with the decision maker. He makes his judgement on the basis of one trade-off ratio at a time, given the corresponding levels of attainment of all of the objectives.

When the decision vector $\underline{x}$ has a great many components and there are three or more non-commensurate objectives the concepts described above still apply.   Methods will be developed in chapters six and seven for computing the trade-off ratios significant to the analysis with a minimum of computational cost.

The approach to the computations in the general problem described in subsequent sections begins by finding the minimum value of each objective function subject to the system of constraints. The minimum of the ith objective function, ignoring all other (n-1) objectives, is determined as in section 1.4 and denoted by $\bar{f}_i$. The next step is to formulate the multiobjective problem in the ε-constraint form (as discussed in section 2.6); the maximum tolerable levels ε_j will be related to $\bar{f}_j$ in the next section.

By considering one objective function as primary and all others as constraints at minimum satisficing levels, the Lagrange multipliers related to the other (n-1) objectives will be zero or non-zero. If non-zero, that particular constraint does limit the optimum. It will be shown that non-zero Lagrange multipliers correspond to the non-inferior set of solutions, while the zero Lagrange multipliers correspond to the inferior set of solutions. Furthermore, the set of non-zero Lagrange multipliers represent the set of trade-off ratios between the principal objective and each of the constraining objectives respectively. Clearly, these Lagrange multipliers are functions of the optimal level attained by the principal objective function as well as of the levels of all objectives satisfied as equality (binding) constraints. Consequently, these Lagrange multipliers form a matrix of trade-off rate functions.

3.2   THE DERIVATION OF THE TRADE-OFF RATE FUNCTION

Given the multiobjective problem posed in problem 1-1, the trade-off rate function between the ith and j^{th} functions denoted by $T_{ij}(\underline{x})$ is defined as follows:

$$T_{ij}(\underline{x}) = \frac{df_i(\underline{x})}{df_j(\underline{x})}$$

where

$$df_i(\underline{x}) = \sum_{k=1}^{N} \frac{\partial f_i(\underline{x})}{\partial x_k} \, dx_k$$

or equivalently,

$$T_{ij}(\underline{x}) = \frac{(\nabla_{\underline{x}} f_i(\underline{x}) \, , \, d\underline{x})}{(\nabla_{\underline{x}} f_j(\underline{x}) \, , \, d\underline{x})} \tag{1}$$

The functions $T_{ij}(\underline{x})$ have the property that

$$T_{ij}(\underline{x}) = 1 \quad , \quad \text{for } i = j$$

and

$$T_{ij}(\underline{x}) = \frac{1}{T_{ji}(\underline{x})} \quad , \quad \text{for all } i \text{ and } j$$

More will be said on the properties of the trade-off rate functions $T_{ij}(\underline{x})$ in subsequent sections. The derivation and determination of the functions $T_{ij}(\underline{x})$ is of primary importance in the SWT method. The direct utilization of equation (1) however, is clearly impractical and computationally prohibitive. Thus, an alternative approach must be sought. The concept of duality and Lagrange multipliers as well as the ε-constraint approach are utilized in subsequent sections of this book to provide both the information needed and a basis for constructing the trade-off rate matrix.[2]

The following development shows that the trade-off rate functions can be found from the values of the dual variables associated with the constraints in a reformulated problem.

Reformulate the system in problem 1-1 as follows:

Problem 3-1: $\min\limits_{\underline{x}} f_1(\underline{x})$ subject to

$$f_j(\underline{x}) \leq \varepsilon_j \ , \quad j = 2, 3, \ldots, n$$

and

$$g_k(\underline{x}) \leq 0 \ ; \quad k = 1, 2, \ldots, m$$

where

$$\varepsilon_j = \bar{f}_j + \bar{\varepsilon}_j, \quad j = 2, 3, \ldots, n$$

$$\bar{\varepsilon}_j > 0 \qquad j = 2, 3, \ldots, n$$

Note that ε_j is defined in terms of $\bar{f}_j$, the minimum value of the jth objective when all other objectives are ignored (see section 1.4), and that the $\bar{\varepsilon}_j$ are the deviations from this minimum value. Thus $\bar{\varepsilon}_j$ represents the value ε_j in the objective space whose f_j axis is shifted to $\bar{f}_j$. The values of $\bar{\varepsilon}_j$ will be varied parametrically in the process of constructing the trade-off functions.

Form the generalized Lagrangian, L , to problem 3-1:

$$L = f_1(\underline{x}) + \sum_{k=1}^{m} \mu_k \, g_k(\underline{x}) + \sum_{j=2}^{n} \lambda_{1j}(f_j(\underline{x}) - \varepsilon_j) \qquad (2)$$

where μ_k, $k = 1,2, \ldots, m$, and λ_{1j}, $j = 2,3, \ldots, n$ are generalized Lagrange multipliers. The subscript $1j$ in λ denotes that λ is the Lagrange multiplier associated (in the ε-constraint vector optimization problem) with the first objective and the jth constraint. The Lagrange multiplier λ_{1j} will be subsequently generalized to be λ_{ij} associated with the ith objective function and the jth constraint.

In order to limit the derivation of the Kuhn-Tucker conditions in this book, denote by χ the set of all x_i, $i = 1,2, \ldots, N$, which satisfy the Kuhn-Tucker conditions in problem 3-1. Similarly, let Ω be the set of all Lagrange multipliers which satisfy the Kuhn-Tucker conditions. For stationary values of $\underline{x}$, μ_k, and λ_{1j} ($k = 1,2, \ldots, m$; $j = 2,3, \ldots, n$), the Kuhn Tucker[3] conditions of interest to our analysis are:

$$\lambda_{1j}(f_j(\underline{x}) - \varepsilon_j) = 0 \quad , \quad j = 2,3, \ldots, n \qquad (3a)$$

$$\lambda_{1j} \geq 0 \quad , \quad j = 2,3, \ldots, n \qquad (3b)$$

Clearly, equation (3a) holds only if $\lambda_{1j} = 0$, or $f_j(\underline{x}) - \varepsilon_j = 0$, or both. Note, however that if $f_j(\underline{x}) - \varepsilon_j < 0$ for any $j = 2,3, \ldots, n$ then the corresponding $\lambda_{1j} = 0$. For the case where the jth constraint is inactive (not binding), the corresponding Lagrange Multiplier (dual variable or shadow price) is identically zero. The set of inactive (non-binding) constraints associated with a specific value of ε_j will be denoted by $I(\varepsilon_j)$

$$I(\varepsilon_j) = \{j : \underline{x} \in \chi \, ; \, f_j(\underline{x}) - \varepsilon_j < 0 \, ; \, j = 2,3, \ldots, n\}$$

Denote the set of active (binding) constraints associated with a specific value of ε_j by $A(\varepsilon_j)$:

$$A(\varepsilon_j) = \{j : \underline{x} \in \chi \, ; \, f_j(\underline{x}) - \varepsilon_j = 0 \, ; \, j = 2,3, \ldots, n\}$$

From equation (3a), it is clear that all λ_{1j} corresponding to ε_j for $j \in I(\varepsilon_j)$ are identically zero. In addition, all λ_{1j} corresponding to ε_j for $j \in A(\varepsilon_j)$ are non-negative and not necessarily zero. Denote these λ_{1j} by $\lambda_{1j}(A(\varepsilon_j))$.

The value of $\lambda_{1j}(A(\varepsilon_j))$, $j = 2,3 \ldots, n$ is of special interest since it indicates the marginal benefit (cost) of the objective function $f_1(\underline{x})$ due to an additional unit of ε_j. From equation (2), the following is derived[4]:

$$\lambda_{1j}(\varepsilon_j) = - \frac{\partial L}{\partial \varepsilon_j} \quad ; \quad j = 2,3, \ldots, n$$

Note, however, that for $\underline{x} \varepsilon \chi$ and $\lambda_{1j} \varepsilon \Omega$, $\mu_k \varepsilon \Omega$, for all j and k,

$$f_1(\underline{x}) = L$$

thus

$$\lambda_{1j}(\varepsilon_j) = - \frac{\partial f_1(\underline{x})}{\partial \varepsilon_j} \quad ; \quad j = 2,3, \ldots, n$$

Also note that for all $\lambda_{1j}(A(\varepsilon_j))$, $f_j(\underline{x}) = \varepsilon_j$, $j = 2,3, \ldots, n$ (since these constraints are active). Therefore:

$$\lambda_{1j}(A(\varepsilon_j)) = - \frac{\partial f_1(\underline{x})}{\partial f_j(\underline{x})}$$

Clearly, this equation can be generalized where the index of performance is the ith objective function of problem 1-1 rather than the first one. In this case the index i should replace the index 1 in λ_{1j} yielding λ_{ij}.

Accordingly:

$$\lambda_{ij}(A(\varepsilon_j)) = - \frac{\partial f_i(\underline{x})}{\partial f_j(\underline{x})} \tag{4}$$

$$i \neq j \quad ; \quad i,j = 1,2, \ldots, n$$

Thus T_{ij} or $\frac{\partial f_i(\underline{x})}{\partial f_j(\underline{x})}$ can be found by calculating $-\lambda_{ij}$ which is obtainable from the overall system Lagrangian as will be discussed subsequently. It is important to note that the trade-off rate function $\lambda_{ij}(A(\varepsilon_j))$ is applicable to any noncommensurable functions. For example, let the units of $f_i(\underline{x})$ be \$, and the units of $f_j(\underline{x})$ be pounds of DO (dissolved oxygen). Then the units of λ_{ij} are \$/DO.

Equation (4) is valid for all $\lambda_{ij}(A(\varepsilon_j))$; i.e., the trade-off ratio is valid only when the jth constraint is active (binding). It can be shown that a direct correspondence exists between $\lambda_{ij}(A(\varepsilon_j))$ (λ_{ij} associated with the active constraints) and the non-inferior set to problem 1-1, and between $\lambda_{ij}(I(\varepsilon_j))$ (λ_{ij} associated with the inactive constraints) and the inferior set. Consider the case where for some ε_j the corresponding λ_{ij} is zero, i.e., $\lambda_{ij} = \lambda_{ij}(I(\varepsilon_j))$. Except for the degenerate case, this means that there is no improvement in the objective function $f_i(\underline{x})$ even at the expense of further degradation of the objective $f_j(\underline{x})$. This solution clearly belongs to the inferior set. The degenerate case where the Lagrange multiplier corresponding to an active constraint is zero has been studied[5]. The solution corresponding to such $\lambda_{ij}(A(\varepsilon_j)) = 0$ is defined here to be associated with the inferior set; i.e., degenerate solutions are considered as inferior solutions.

Consider next the case where for some ε_j the corresponding $\lambda_{ij} = \lambda_{ij}(A(\varepsilon_j))$. This means that $\lambda_{ij} > 0$, i.e., there is a degradation in the jth objective function for an improvement in the ith objective function

since $\lambda_{ij} = -\dfrac{\partial f_i(\underline{x})}{\partial f_j(\underline{x})}$. Thus, this solution corresponds to the non-inferior set.

In summary, since only the non-inferior solutions are of interest, only $\lambda_{ij}(A(\varepsilon_j)) > 0$ need be considered. For simplicity in notation, the letter A, indicating active constraint, will be dropped and the symbol $\lambda_{ij}(\varepsilon_j)$ will be used hereafter.

The possible existence of a duality gap[6] and its effect on the SWT method is discussed in chapter 4. Note that even if a duality gap does exist, the ε-constraint method still generates all non-inferior solutions. However, a given value of the trade-off rate function λ_{ij} , may correspond to more than one non-inferior solution.

3.3 COMPUTATIONAL PROCEDURE FOR CONSTRUCTING THE TRADE-OFF FUNCTION

In this section, a possible approach is presented for generating the trade-off ratios $T_{ij} = \partial f_i/\partial f_j$ by calculating the Lagrange multipliers λ_{ij} to problem 3-1. First λ_{12} will be found as a function of ε_2 .

The system given by problem 3-1 is solved for k values of ε_2 , say $\varepsilon_2^1, \varepsilon_2^2, \ldots, \varepsilon_2^K$, where all other ε_j , j = 3,4, ..., n are held fixed at some level ε_j^0 . Of course, only the positive $\lambda_{12}(\varepsilon_2^k)$, k = 1,3, ..., K, are of interest (corresponding to the non-inferior solutions). Since these $\lambda_{12}(\epsilon_2^K)$ are positive, it must be that $f_2^k(\underline{x}) = \epsilon_2^k$ (active constraints), for k = 1,2, ..., K. Thus, for each value of $f_2^k(\underline{x})$ (k = 1,2, ..., K) where the constraints are active, a corresponding functional value of $\lambda_{12}(f_2^k(\underline{x}))$ (k = 1,2, ..., K) is generated. At this stage a regression analysis may be performed if desired, to yield a least squares approximation to the function $\lambda_{12}(f_2(\underline{x}))$ (figure 3-1 depicts a quadratic function fit by regression). Note, however, that $\lambda_{12}(f_2(\underline{x}))$ is also a function of the values of ϵ_j , j = 3,4, ..., n. If the function is sensitive to these levels of ϵ_j , j = 3,4, ..., n , then a multiple regression analysis can be performed. It will be shown in Chapter 4 that often the trade-off rate function λ_{ij} need be constructed only in the vicinity of the indifference band. Interpolation and curve fitting procedures can also be used to approximate λ_{ij} near the indifference band. This minimizes the computational effort involved in the regression analysis.

Similarly, the trade-off rate function λ_{13} can be generated, where again the prime objective function is $f_1(\underline{x})$, and problem 3-1 is solved for K different values of ϵ_3^k , k = 1,2, ..., K , with a fixed level of $\epsilon_2^o, \epsilon_4^o, ..., \epsilon_n^o$. This technique is repeated to generate the trade-off rate functions λ_{1j}, for j = 4,5, ..., n. Once all trade-off rate functions λ_{1j}, j = 2,3, ..., n, have been generated, the prime objective is changed to the ith and thus all trade-off rate functions λ_{ij}, i≠j , j = 1,2, ...,n can be generated.

When all of the functions λ_{ij} have been determined for all i, j = 1,2, ..., n, i ≠ j, it is sometimes desirable to transform the λ_{ij} into functions of the decision variables $\underline{x} \epsilon \chi$. This transformation exists and can be shown to be unique with certain assumed properties on the $f_j(\underline{x})$ and $\lambda_{ij}(f_j(\underline{x}))$, j = 1,2, ..., n.

Example:

Assume that $f_2(\underline{x})$ is given as linear function of $\underline{x}$:

$$f_2(\underline{x}) = a_1x_1 + a_2x_2 + a_3$$

and that $\lambda_{12}(f_2(\underline{x}))$ is given as a quadratic function of $f_2(\underline{x})$:

$$\lambda_{12}(f_2(\underline{x})) = b_1f_2(\underline{x}) + b_2(f_2(\underline{x}))^2 + b_3$$

Then λ_{12} can be obtained as a function of $\underline{x}$ as follows:

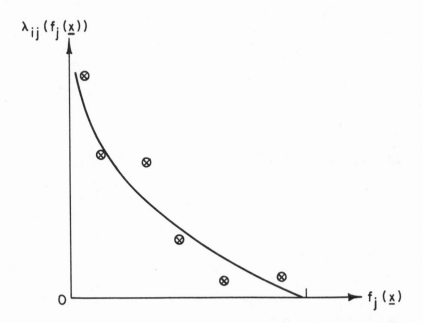

Figure 3-1. Regression Fitting of $\lambda_{ij}(f_j(\underline{x}))$ as a function of $f_j(\underline{X})$

$$\lambda_{12}(\underline{x}) = c_1 x_1 + c_2 x_2 + c_3 x_1 x_2 + c_4 x_1^2 + c_5 x_2^2 + c_6$$

$$(x_1, x_2) \ \varepsilon \ \chi$$

Theoretically the following relations should hold:

$$\lambda_{ij}(\underline{x}) = \frac{1}{\lambda_{ji}(\underline{x})} \quad ; \quad i,j = 1,2, \ldots, n$$

For computational advantage this property can be used to check the consistency of the regression functions and to ensure that no excessive regression functional deviations occur.

3.4   THE SURROGATE WORTH FUNCTION

Assume that a particular value of the trade-off ratio λ_{ij}, $i \neq j$, between two objectives $f_i(\underline{x})$ and $f_j(\underline{x})$ has been computed. If this particular value were by chance the preferred solution, then to the decision maker, λ_{ij} units of $f_i(\underline{x})$ would be exactly equal in value to one unit of $f_j(\underline{x})$. The term preferred solution was defined to be any non-inferior feasible solution belonging to a subset of the non-inferior set - the indifference band - where the worth of an improvement in one objective is equivalent to the corresponding degradation in another. Since the attained levels of all objectives are known, it should be relatively simple to answer this question: Is the given marginal change (λ_{ij}) in the ith objective function $(f_i(\underline{x}))$ worth more or less than a one unit change in the jth objective function $(f_j(\underline{x}))$? If the first alternative is believed to be true, then objective $f_i(\underline{x})$ should be decreased at the expense of $f_j(x)$, and vice versa. In either case the desire is to locate the particular set of trade-off ratios λ_{ij} which are simultaneously neutral to this question.

A surrogate worth function W_{ij}, $i \neq j$; $j = 1,2, \ldots, n$, can be defined as any monotonic function of λ_{ij} estimating the desirability of the trade-off λ_{ij}. For example, the scale could range from - 10 to +10, with a -10 indicating that λ_{ij} marginal units of objective i are much less desirable than an additional unit of j , a +10 indicating the opposite, and a zero signifying an even trade (i.e., belongs to the indifference band) A similar surrogate worth function can be defined for each trade-off ratio. The preferred solution is where the selected trade-off ratios make the surrogate worth functions simultaneously equal to 0 (or such other number as may be designated as the measure of an even trade).

In practice there will usually be a band of indifference of attained objectives near any preferred solution thus identified. That is, there will be a range of values of f_i and f_j over which the decision maker would not be prepared to assert that Δf_i is worth more or less than Δf_j . By approaching the zero using only the positive limb of the surrogate worth function one bound of the band of indifference is established. By approaching zero using only the negative limb, the opposite bound is determined, thus establishing the band of indifference.

Any surrogate worth function can be used so long as it has a specified value at indifference (e.g., zero) and is monotonic. For example, the function could arbitrarily be assigned a value of +10 for the first trial non-inferior solution if the worth of $\Delta f_i > \Delta f_j$, and a value of -10 if the worth of $\Delta f_i < \Delta f_j$. By maintaining monotonic consistency, the form of a satisfactory surrogate worth function can be determined by successive trials of non-inferior solutions. In particular, earlier trials may be used to estimate where the zero might be to accelerate movement towards the indifference band.

If the process were repeated using different starting points, or if different positive (or negative) values were assigned at intermediate steps, different surrogate worth functions would be generated. However, by definition of the zero of these functions, the same solution (or band of indifference) would result.

The uniqueness of the resulting band of indifference permits application of the surrogate worth method to situations where there is more than one decision maker. If all have the same standards of value, the same decision (or band of indifference) will be determined regardless of the surrogate functions developed. If the plural decision makers have different standards of value, as would normally be expected, the surrogate worth method has the property of defining the indifference band at any and all levels of unanimity or lack thereof. The band of indifference for unanimity would be that set of decisions including all decision makers' bands of indifference. By defining nested intervals, each containing one less decision maker's band of indifference, bands of indifference for less than unaminious majorities can be approximated. Approximation stems from the fact that some rule will have to be devised to determine which band of indifference should be deleted next. An obvious rule might be to specify that candidate which would narrow the group band of indifference by the greatest amount. Alternatively, a specific decision could be "defined" as preferred if it is included in the maximum number of individual indifference bands. Of course, all such rules are somewhat arbitrary even when they seem intuitively reasonable or appeal to some logic of social justice.

In summary, the surrogate worth function W_{ij} associated with the ith and j^{th} objectives can be defined as any monotonic function satisfying:

$$
W_{ij} \begin{cases}
> 0 & \text{when } \lambda_{ij} \text{ marginal units of } f_i(\underline{x}) \text{ are preferred over one} \\
& \text{marginal unit of } f_j(\underline{x}), \text{ given the level of achievement} \\
& \text{of all the objectives.} \\
\\
= 0 & \text{when } \lambda_{ij} \text{ marginal units of } f_i(\underline{x}) \text{ are equivalent to one} \\
& \text{marginal unit of } f_j(\underline{x}), \text{ given the level of achievement} \\
& \text{of all the objectives.} \\
\\
< 0 & \text{when } \lambda_{ij} \text{ marginal units of } f_i(\underline{x}) \text{ are not preferred over} \\
& \text{one marginal unit of } f_j(\underline{x}), \text{ given the level of achieve-} \\
& \text{ment of all the objectives.}
\end{cases}
$$

It is important to note here that the decision maker(s) is provided

with the trade-off value (via the trade-off rate function) of any two objective functions at a stated level of attainment of all of the objective
functions. Furthermore, all trade-off values generated from the trade-off
function are associated with the non-inferior set.   Hence, it is evident
that it is always possible to generate a surrogate worth function which will
in turn determine the band of indifference of λ_{ij} ($i{\neq}j$, $i,j = 1,2, \ldots, n$)
and thus the preferred solutions to the multi-objective problem (the computational procedure for finding the overall solution to problem 3-1 from the
surrogate worth functions is discussed in the next section).

The specific values of the surrogate worth W_{ij} assigned by the DM
for each value of λ_{ij} are unimportant, needing only to be monotonic in λ_{ij}
and to pass through zero at indifference.   Thus W_{ij} are ordinal measures,
compared to the cardinal measures required by the utility function approach.

Finally, it is important to note that the above analyses are conduc-
ted in the functional space, $f_1(\underline{x}), \ldots, f_n(\underline{x})$, and not in the decision
space, $x_1, \ldots, x_N$. This is of course, a clear advantage since typically
$N \gg n$ (one may have a hundred or a thousand decisions even with just three
to ten objective functions).

3.5   COMPUTATIONAL PROCEDURE FOR FINDING THE PREFERRED SOLUTIONS

The surrogate worth function described in the previous   section
assigns a scalar value (on an ordinal scale) to any non-inferior   solution
to the multiple objective problem 1-1. However, there are three different
ways of specifying a non-inferior solution - by its decision space coordinates, by its objective function space coordinates, or by its trade-off rate
values λ_{ij} at the non-inferior point. Thus, there are three possibilities
for the surrogate worth function $W_{ij}(x_j)$, $W_{ij}(f_j)$ or $W_{ij}(\lambda_{ij})$.

3.5.1   Decision Space Surrogate Worth Function

In this approach, the surrogate worth function is developed as a
function of the decisions $x_1 \ldots x_N$; however, there are several reasons why
this is generally not feasible. First, the values of the decisions are not
particularly relevant or meaningful to the DM , since he is more familiar
(and more directly concerned) with the objectives and their trade-off values.
Second, the decision space is generally more complicated than the others
since there are usually many more decisions than objectives. Third, one must
restrict the values of the decisions to non-inferior values. In the functional space this can be easily done since any n-1 of the objectives specify
a non-inferior point via the trade-off function. This is impossible in the
decision space, however (see figure 1-1). Thus the problem of finding a

point where both $W_{ij}(\underline{x}) = 0$ and $\underline{x}$ is non-inferior makes this approach undesirable.

3.5.2 λ-Space Surrogate Worth Function

In this approach, the surrogate worth function W_{ij} is developed as a function of λ_{ij}. In general, values of λ_{ij} for n-1 different j are required to specify a non-inferior point, but for the present, this approximation will be maintained. For several distinct values of λ_{ij}, the decision maker is asked whether λ_{ij} units of $f_i(\underline{x})$ are more or less or equally preferred to one unit of $f_j(\underline{x})$. A linear combination of the two answers $W_{ij}(\lambda_{ij})$ nearest zero can be made (see figure 3-2). Then the value of $\lambda_{ij} = \hat{\lambda}_{ij}$ is chosen such that $W_{ij}(\hat{\lambda}_{ij}) = 0$ on the line segment fitting the two values of λ_{ij}. With this estimate $\hat{\lambda}_{ij}$, the corresponding $W_{ij}(\hat{\lambda}_{ij})$ is requested and the process repeated until indifference is found at λ_{ij}^*. Then the indifference band is assumed to exist within a neighborhood of λ_{ij}^*. Additional questions to the decision maker can be asked in the neighborhood of λ_{ij}^* to determine the approximate limits of the band of indifference. Having determine all λ_{ij}^* for i, j = 1,2, ..., n , the following system of relations should be solved simultaneously in the decision space $\underline{x}$:

Problem 3-2:

Solve: $\lambda_{ij}(\underline{x}) = \lambda_{ij}^* $; i,j = 1,2, ..., n

                                    i ≠ j

such that $\underline{x} \in \chi$

This problem represents n^2-n equations with the constraints derived from the Kuhn-Tucker conditions that should be imposed on $\underline{x}$. Note that since normally N >> n, the number of equations in problem 3-2, (n^2-n) may still be much smaller than N. The constraints on $\underline{x}$ introduce additional inequalities which in turn limit the solution space to only feasible non-inferior solutions. A non-unique preferred (indifference) solution can be expected and is of course, accepted, as in any optimization problem with a single objective function. For the case where $\lambda_{ij}(f_j(\underline{x}))$ is constant over a given variation in $f_j(\underline{x})$, as may happen with linear objective functions, an elasticity term can be augmented to λ_{ij} for determining $W_{ij}(\lambda_{ij})$. The trade-off rate function λ_{ij} can be replaced by the elasticity trade-off rate function.

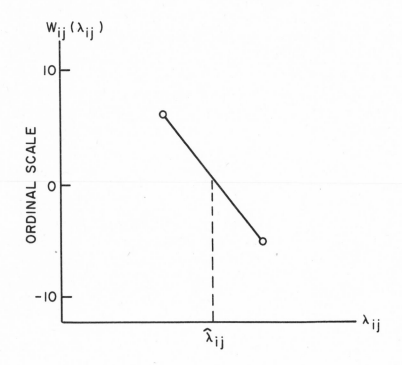

Figure 3-2.   Determination of the indifference band

$$\lambda'_{ij} = - \frac{f_j(\underline{x})}{f_i(\underline{x})} \frac{\partial f_i(\underline{x})}{\partial f_j(\underline{x})}$$

Alternative elasticity functions may be constructed to circumvent this problem. Other techniques are described in the next section.

Although this approach appears to suffer from the problem of insuring that $\underline{x}$ is non-inferior when reverting to the decision space, a means
of avoiding this will be discussed in sections 4.1.3 and 6.3.2 .   A more
severe problem, however, is that this approach is guaranteed only when the
trade-off function is convex and non-linear.

Consider the two objective static case where the trade-off f_1 vs. f_2 is non-convex as shown in figure 3-3-a.   In such a case, it is obvious that $\lambda_{12}(f_2) = -df_1/df_2$ is not a unique function over the range of values of f_2 in the non-inferior region (shown in figure 3-3-b). (Note that the total derivative is present since there are only two objectives.)   Hence, worth is no longer a well defined function of λ_{12}, since for some values of λ_{12} there will be more than one corresponding point (f_1, f_2) on the trade-off curve, and in general, a different worth will be assigned to each point. Thus   the worth may appear as in figure 3-3-c. Of course, in some cases W_{12} may   be independent of f_1 and f_2, so that $W_{12}(\lambda_{12})$ will be a well defined   function (as shown in figure 3-3-d).

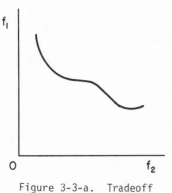

Figure 3-3-a. Tradeoff
              Function

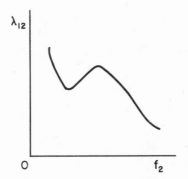

Figure 3-3-b. Tradeoff Rate
              Function

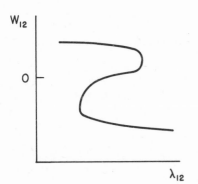

Figure 3-3-c. Possible Worth
              Curve

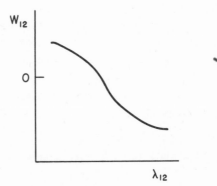

Figure 3-3-d. Possible Worth
              Curve

Figure 3-3. λ-Space Worth Functions for Non-Convex Problems

When worth is not a well defined function of λ_{ij} however, this tends to cause two types of problems. First, if the worth is found only at certain values of λ_{12} and a polynomial fit is attempted, the solution to $W_{12}(\lambda_{12}) = 0$ will probably be quite different from the true preferred trade-off rate λ_{12}^*. Second, assume that enough information is obtained in the first attempt to find the actual shape of the multi-valued worth curve (in general this will be computationally infeasible); then the true preferred trade-off rate λ_{12}^* can be determined. However, there may be several points on the trade-off curve which have the same slope of $-\lambda_{12}^*$. Thus, when λ_{12}^* is related back to the system model to find the preferred decision vector, the model will not be able to discern which of these points is the preferred solution.

The other pathological case is where $\lambda_{12}(f_2)$ is constant over a certain interval (see figures 3-4-a and 3-4-b). This always occurs in linear problems and sometimes in non-linear ones. Since λ_{12} is again not one-to-one there will be several values of the worth corresponding to some values of λ_{12}. Furthermore, in the linear problem, since $\lambda_{12}(f_2)$ is also discontinuous there are no values of the worth for other values of λ_{12} (see figure 3-4-c). Thus, incorrect or even infeasible results for λ_{12}^* may be found. Earlier in this section, it was suggested to use a transformation $\lambda_{12}' = f_2$ λ_{12}/f_1 to circumvent the linearity of $\lambda_{12}(f_2)$. The rationale is to take the elasticity into account; there are still two problems present with this approach, however:

(1) $\lambda_{12}'(f_2)$ may still be discontinuous and thus the worth function will also be discontinuous, again with possibly infeasible results; note however that it may be possible to circumvent this problem by some sort of regression over these intervals since the discontinuities are bounded.

(2) The geometric interpretation is now lost since λ_{12}' is no longer the negative of the slope of the trade-off curve - and the definition of the preferred solution must be changed.

Although the λ-space approach appears to have many drawbacks, it is the simplest for some problems. Algorithms using this approach are included in subsequent chapters.

3.5.3   Objective Function Space Surrogate Worth Function

A simple means of surmounting these difficulties is to use the objective function space surrogate worth function. This will be developed here for the two-objective case; the generalization for n objectives is provided in chapter 6. Note that in general the values of n-1 of the objective are

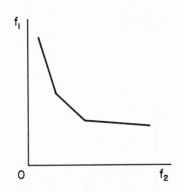

Figure 3-4-a.   Tradeoff Function

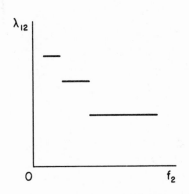

Figure 3-4-b.   Tradeoff Rate
Function

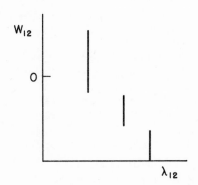

Figure 3-4-c   Possible Worth Curve

Figure 3-4. λ-Space Worth Functions for Linear Problems.

needed to specify a non-inferior point, so that W_{ij} will be a function of n-1 of the f_j.

For the two-objective problem, the first segment of the SWT method provides the functions $f_1^*(f_2)$ and the corresponding $\lambda_{12}(f_2)$ $=$ $- df_1^*/df_2$. Using the same definition of worth as previously, the surrogate worth function tion W_{12} can be developed as a function of f_2 directly, and $W_{12}(f_2)= 0$ solved to find f_2^* (the preferred value of f_2).   The preferred trade-off   rate

λ_{12}^* is $\lambda_{12}(f_2^*)$ and the preferred value of f_1 is $f_1^*(f_2^*)$.

For any value $\hat{f}_2$, the decision maker is still asked how much $\lambda_{12}(\hat{f}_2)$ additional units of f_1 are worth relative to one additional unit of f_2, given $\hat{f}_2$ units of f_2 and $f_1^*(\hat{f}_2)$ units of f_1 . His relative assessment, say, on the scale from - 10 to + 10, is the value $W_{12}(\hat{f}_2)$. By asking enough questions at various values of f_2, points can be generated so that the functions $W_{12}(f_2)$ can be approximated (e.g., by regression).

Since for each value of f_2 there is only one value of λ_{12} and f_1 in the non-inferior region (i.e., λ_{12} and f_1 are both functions of f_2 even though f_2 may not be a function of λ_{12}), $W_{12}(f_2)$ will be a unique function of f_2, even for the non-convex and linear cases causing the previous diffi- culties. Consider the case where $\lambda_{12}(f_2)$ was constant over a certain inter- val. This new approach implicitly takes elasticity into account since one can find the worth at different values of f_2 in the interval where $\lambda_{12}(f_2)$ is constant. Also note that this approach allows the decision maker to de- termine the actual elasticity rather than using any preconceived value such as f_2/f_1.

In many problems, there may be more than one solution to $W_{12}(f_2)=0$; each of these is a preferred solution (the social indifference curve is tan- gent to the trade-off curve at more than one point) and some other criterion is necessary to choose among them.

Furthermore, with this approach the transformation to the decision space can be accomplished without worrying about non-inferiority. The solu- tions to $W_{12}(f_2) = 0$ are the preferred values f_2^* . Accordingly, the prefer- red vector decisions $\underline{x}^*$, can be obtained by simply solving the following optimization problem :

Problem 3-3:

$$\min_{\underline{x}} f_1(\underline{x})$$

$$\text{subject to} \quad f_2(\underline{x}) \leq f_2^*$$

$$g_k(\underline{x}) \leq 0 \ , \ k = 1,2, \ \dots, \ m$$

This is just a common problem in single objective optimization; it is the same as problem 3-1 with ε_j replaced by f_j^*. Its solution yields the desired $\underline{x}^*$ for the total vector optimization problem 1-1. Note that this alterna- tive method avoids the need of representing λ_{ij} as a function of $\underline{x}$ (e.g.,

$\lambda_{ij}(\underline{x}))$, and consequently eliminates the need of solving problem 3-2.

Thus it can be seen that the objective function space is generally the best domain for the surrogate worth function. Algorithms utilizing this approach are presented in the following chapters.

3.6 GEOMETRIC INTERPRETATION OF THE SWT METHOD

The case of two objective functions is considered first; this will be later generalized to n objectives. The first segment of the SWT method develops points on the trade-off curve $f_1^*(f_2)$ and trade-off rate curve $\lambda_{12}(f_2)$ at various values of f_2. The decision maker is then questioned to determine the worth corresponding to various non-inferior values of f_2; he is asked to specify the relative worth (on a scale of (say) +10 to -10) of $\lambda_{12}(f_2)$ additional units of f_1 compared to one additional unit of f_2, given $f_1^*(\hat{f}_2)$ and $\hat{f}_2$ units of the two objectives. His assessment of this relative worth indicates the divergence between the negative slope of the (or his) indifference curve (m) and λ_{12}, at the point $(f_1^*(\hat{f}_2),\hat{f}_2)$ in the functional space.

Consider figure 3-5. The Social Indifference (SI) curves representing equal preference are in reality unavailable; however, it is assumed that the DM is basing his decisions on his subjective ideas of their form. The trade-off curve T is also generally impossible to determine completely; usually only a finite number of points on the curve will be known. At point A, for example, the DM is willing to accept only m^A additional units of f_1 to get one unit less of f_2. However $\lambda_{12}^A > m^A$ so that when asked if he is willing to take λ_{12}^A additional units of f_1 for one less of f_2 his response will be "no", thus $W_{12}(\lambda_{12}^A) < 0$. At point C, $m^C > \lambda_{12}^C$ so that when asked if he is willing to take λ_{12}^C additional units of f_1 for one less of f_2, his response will be "yes", or $W_{12}(\lambda_{12}^C) > 0$. At point B, $m^B = \lambda_{12}^B$ so that when asked if he is willing to take on λ_{12}^B additional units of f_1 for one less of f_2, his response will be indifference, or $W_{12}(\lambda_{12}^B) = 0$. Note that the actual numerical value of the worth functions represents only a relative scaling: if $W_{12}(\lambda_{12}^0) = +8$ and $W_{12}(\lambda_{12}^1) = +7$, all that can be said is that the difference between m and λ_{12} at the point on the trade-off curve corresponding to λ_{12}^0 is greater than the difference between them at the point corresponding to λ_{12}^1. These numerical values can be used, however, as a first estimate of λ_{12}^*; e.g., λ_{12}^* is seven times farther from λ_{12}^1 than λ_{12}^1 is from λ_{12}^0.

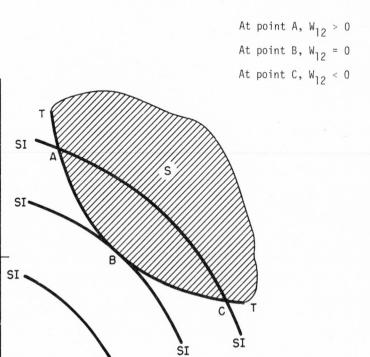

At point A, $W_{12} > 0$

At point B, $W_{12} = 0$

At point C, $W_{12} < 0$

S is the feasible set in the functional space

The thick boundary of S is the trade-off function T

SI are the social indifference curves

at A, the slope of T is $- \lambda_{12}^{A}$, the slope of SI is $-m^{A}$

at B, the slope of T is $- \lambda_{12}^{B}$, the slope of SI is $-m^{B}$

at C, the slope of T is $- \lambda_{12}^{C}$, the slope of SI is $-m^{C}$

Figure 3-5. Geometric Interpretation of Worth Function.

3.7 SUMMARY

The surrogate worth trade-off method provides a means of finding the preferred solution (i.e., maximum utility) by determining the point of tangency between the trade-off function and the social indifference curve. The trade-off rate functions enable the decision maker to compare the slopes of the trade-off function and social indifference curve at various points in the functional space; these trade-off rate functions were found to be the Lagrange multipliers of the multiple objective problem in ε-constraint form. The surrogate worth function allows interpolation of the DM's responses to find the preferred solution.

The SWT method can be viewed as an intermediary between the DM and the system response (figure 3-6), both simplifying and quantifying their interaction. The SWT method initially interacts with the system to determine the trade-off and trade-off rate functions among the objectives. It then interacts with the DM by developing questions of relative worth for his assessment. By analyzing his ordinal responses the preferred solution is found. This solution (via the SWT method) can then be expressed in terms of the decision variables of the system. There is still considerable personal interaction between the DM and the system since his response will normally reflect his knowledge of the system and his evaluation of the preferences of his constituents. The SWT method, however, greatly simplifies the DM's task because many of the non-inferior solutions can be systematically eliminated based on (and fully compatible with) his knowledge and expressed preferences.

The advantages of the SWT method are numerous. The decisions required by the DM are minimal; he deals only with the functional space (which is generally much smaller, easier to work with, and of direct rather than indirect significance when compared to the decision space), and with ordinal relationships between his values (rather than actual values). No assumption is made concerning the form of the utility function; only that indifference by the DM to a particular trade-off properly represents indifference on the part of his constituents. Of course his value judgements are necessarily subjective, but he is provided with adequate information, and a logical framework, to rationally and simply assess and evaluate his preferences.

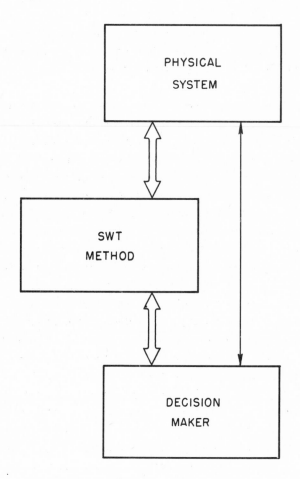

Figure 3-6.   Role of the SWT Method.

FOOTNOTES

1. The original development of this method can be found in Haimes and Hall [1974].

2. The concepts of duality and Lagrange multipliers are discussed in Everett [1963], Lasdon [1968] and others, while the ε-constraint approach was discussed in section 2.6 (also see Haimes [1973]).

3. These are the necessary conditions for stationarity and can be found in Kuhn and Tucker [1950] as well as in most optimization texts.

4. See Luenberger [1973].

5. For other results on the degenerate case see Olagundoye [1971].

6. Again see Everett [1963], Lasdon [1968], or Gembicki [1973].

REFERENCES

1. Everett, H., "Generalized Lagrange Multipliers Method for Solving Problems of Optimum Allocation of Resources," Operations Research, vol. 11, pp. 399-417, 1963.

2. Gembicki, F., "Vector Optimization for Control with Performance and Parameter Sensitivity Indices," Ph.D. Dissertation, Case Western Reserve University, 1973.

3. Haimes, Y.Y., "The Integrated System Identification and Optimization," in Advances in Control Systems Theory and Applications , C. T. Leondes, Editor, Volume X, Academic Press, N.Y. pp.435-518, 1973 .

4. Haimes, Y. Y., and Hall, W. A., "Multiobjectives in Water Resources Systems Analysis: The Surrogate Worth Tradeoff Method," Water Resources Research, vol. 10, no. 4, pp. 615-624, 1974.

5. Kuhn, H. W., and Tucker, A. W., "Nonlinear Programming," in Second Berkeley Symposium on Mathematical Statistics and Probability, University of California Press, Berkeley, California, 1950.

6. Lasdon, L.S., "Duality and Decomposition in Mathematical Programming," IEEE Transactions, vol. SSC-4, no. 2, 1968.

7. Luenberger, D. G., Introduction to Linear and Nonlinear Programming, Addision-Wesley Publishing Company, Inc., 1973.

8. Olagundoye, O. B., "Efficiency and the ε-Constraint Approach for Multi-Criterion System," M.S. Thesis, Systems Engineering Department, Case Western Reserve University, Cleveland, Ohio, 1971.

Chapter 4

THE SWT METHOD FOR STATIC TWO-OBJECTIVE PROBLEMS

The overall procedure for solving multiple objective problems with
the surrogate worth trade-off method can be divided into two segments. The
first part involves the development of information about the trade-off
function in order to edify the decision maker. The second segment uses
the information provided by the DM's choices to find the preferred solution. The solution algorithms presented in this book can be classified
according to the approach used for each segment - ε-constraint (E), multiplier (M), or combined (C) - and the types of problems for which they are
appropriate - static (S) or dynamic (D), two-objective (T) or n-objective
(N).

This chapter will present algorithms for the two-objective static
case. Innovations to improve computational efficiency of the SWT method
are detailed. The applicability of each algorithm is discussed, and sample
problems are included to illustrate their use.

4.1 COMPUTATIONAL EFFICIENCIES

There are several improvements that can be incorporated in the SWT
method for the two-objective case. First, maximum values for ε_2 in the
non-inferior region can be found to lessen the effort wasted by finding
inferior points (non-binding ε_2). Secondly, it will be shown that once
$W_{12}(f_2)$ has been found, W_{21} is redundant, but it can be used as a consistency check. Third, a simple method of reverting to the decision space
will be presented. Finally, it will be shown that by using search techniques and interpolations the entire SWT procedure can be accomplished without resorting to regressions.

4.1.1 Limits on ε_2

The maximum value, f_{2MAX}, for ε_2 in the non-inferior region is
found by solving the following problem:

Problem 4-1:

$$\text{MIN } f_1(\underline{x})$$
$$\text{s.t. } \underline{x} \in T$$

Theorem: If the solution vector to problem 4-1 is $\underline{x}^*$, then the maximum
value for ε_2 is $f_2(\underline{x}^*)$.

Proof: Since $\underline{x}^*$ solves problem 4-1, then for any other feasible vector $\underline{x}_1 \varepsilon T$, $f_1(\underline{x}_1) \geqslant f_1(\underline{x}^*)$. Therefore any $\underline{x}_1$ which also gives $f_2(\underline{x}_1) > f_2(\underline{x}^*)$ must be in the inferior region since a reduction can be obtained in both objectives by using $\underline{x}^*$. Thus the largest value that f_2 can attain in the non-inferior region is $f_2(\underline{x}^*)$. For the case where the solution vector $\underline{x}^*$ to problem 4-1 is not unique, the minimum of the values $f_2(\underline{x}^*)$ is used as the maximum value for ε_2 .

The minimum value for ε_2 is found as in section 1.4 by ignoring all the other objectives . When the non-inferior set is continuous, the ε_2 constraint will always be binding for values of ε_2 in the interval $f_{2MIN} \leqslant \varepsilon_2 \leqslant f_{2MAX}$. As an example of the case where the constraint is not binding, consider the following problem:

$$\text{MIN } f_1 = x - x^3/3$$
$$\text{MIN } f_2 = x$$
$$\text{s.t. } -3 \leqslant x \leqslant 3$$

The objectives are shown in figure 4-1-a; the feasible set S in the functional space is depicted in figure 4-1-b. When the value $\varepsilon_2 = 0$ (which is between $f_{2MIN} = -3$ and $f_{2MAX} = 3$) is used, the solution will be at $x = -1$, $f_2 = -1$, $f_1 = -2/3$ so that the constraint is not binding. In this example every ε_2 in the interval $-1 < \varepsilon_2 < 2$ will be non-binding.

In summary, when using the ε-constraint approach, large numbers of inferior solutions can be automatically eliminated by finding minimum and maximum values for ε_2. In addition, for each solution found by this approach, the ε_2 constraint should be checked to insure that it is binding; if not binding, then the solution is inferior.

4.1.2 Trade-off and Worth Relationships

The first segment of the SWT method provides points $\underline{f}^T = (f_1, f_2)$ in the non-inferior set, and the trade-off rate $\lambda_{12} = -\partial f_1/\partial f_2|_f$. Similarly there is a value of the trade-off rate $\lambda_{21} = -\partial f_2/\partial f_1|_f$. Although the previous chapter considered λ_{21} as a function of f_1 and λ_{12} as a function of f_2, it is obvious that both can be considered as a function of f_2 since in the non-inferior set f_1 is a known function of f_2 (that is $f_1^*(f_2)$).[1] Then for any value of f_2 , $\lambda_{21}(f_2) = 1/\lambda_{12}(f_2)$. Since the

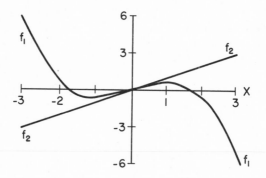

Figure 4-1-a. Objectives in Decision Space.

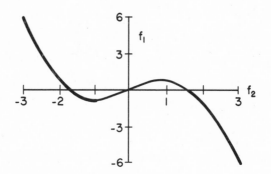

Figure 4-1-b. Functional Space.

Figure 4-1.   Non-Continuous Trade-off Curve.

Note: The feasible set S is the entire curve -
        the non-inferior set is the thick portion.

values $\lambda_{21}(f_2)$ are available, $\lambda_{21}(f_2)$ can be found without resolving the problem with f_2 as the primary objective. It can now be seen that the worth function W_{21} can also be considered a function of f_2; its values are determined by asking the DM how much $\lambda_{21}(\hat{f}_2)$ additional units of f_2 are worth in relation to one additional unit of f_1, given $\hat{f}_2$ units of f_2 and $f_1^*(\hat{f}_2)$ units of f_1. It will be assumed that the trade-off rate λ_{12} is a good approximation to the change which occurs in the non-inferior value of f_1 when f_2 is changed by one unit; that is $|f_2| \gg 1$ and $|f_1| \gg |\lambda_{12}|$. It will be shown in chapter six that if $W_{12}(f_2) = 0$ then $W_{21}(f_2) = 0$, since indifference to trading λ_{12} additional units of f_1 for one additional unit of f_2 is the same as indifference to trading one unit of f_1 for $1/\lambda_{12}$ units of f_2. Thus finding $W_{21}(f_2)$ is redundant. However, decision makers are generally not able to assess their preferences accurately. Since, $W_{21}(f_2)$ can be found at each value f_2 at which W_{12} is found simply by asking the DM one additional question, it may be useful to use an averaged worth function $W_{12}(f_2) = 1/2 \; (W_{12}(f_2) + W_{21}(f_2))$ to find the preferred solution.

4.1.3   Reversion to the Decision Space

The solution of $W_{12}(f_2) = 0$ is the preferred value f_2^* of objective f_2. The preferred decision vector $\underline{x}^*$ can be found as in section 3.5.3 by solving

Problem 4-2:

$$\text{MIN} \quad f_1(\underline{x})$$
$$\text{s.t.} \quad f_2(\underline{x}) \leq f_2^*$$
$$\underline{x} \; \varepsilon \; T$$

It will now be shown that the solution $\underline{x}^*$ to problem 4-2 is the same preferred solution that would be found by the method described in section 3.5.2 if it is modified to remove redundancy. That method is to solve simultaneously:

$$\lambda_{12}(f_2(\underline{x})) = \lambda_{12}^*$$
$$\lambda_{21}(f_1(\underline{x})) = \lambda_{21}^*$$

s.t. $\underline{x}$ satisfies the Kuhn-Tucker conditions ($\underline{x}$ is a non-inferior point). However, $\lambda_{12}(f_2(\underline{x})) = 1/\lambda_{21}(f_1(\underline{x}))$ whenever $\underline{x}$ is a non-inferior point, and $\lambda_{12}^* = 1/\lambda_{21}^*$ if the DM is consistent. Thus the second equation is redundant and this approach becomes:

Problem 4-3:   Solve $\lambda_{12}(f_2(\underline{x})) = \lambda_{12}^*$ such that $\underline{x}$ meets the Kuhn-Tucker conditions for problem 3-1.

Theorem: If $\underline{x}^*$ solves problem 4-2, then it solves problem 4-3.

Proof: Note that there is a value $\lambda_{12}^* > 0$ which is the negative of the slope of the trade-off function at f_2^* .   Assume the function $\lambda_{12}(f_2)$ is known; then $\lambda_{12}^* = \lambda_{12}(f_2^*)$. Since the constraint $f_2(\underline{x}) \leq f_2^*$ is binding (its multiplier is λ_{12}^* which is greater than zero), then $f_2(\underline{x}^*) = f_2^*$. Thus $\lambda_{12}(f_2(\underline{x}^*)) = \lambda_{12}^*$ so $\underline{x}^*$ solves problem 4-3. In addition, $\underline{x}^*$ automatically satisfies the Kuhn-Tucker conditions since it solves the minimization problem 4-2, which is the same as problem 3-1 with ε_2 replaced by f_2^*. Thus this reversion method which does not require knowing $\lambda_{12}(f_2)$ in functional form is much simpler to use.

4.1.4   Regressions

The first segment of the SWT method provides points $\underline{f}^T = (f_1,f_2)$ in the non-inferior set, and the trade-off rate $\lambda_{12} = - \partial f_1/\partial f_2$ evaluated at $\underline{f}$.   For each of these points, one value of the worth function $W_{12}(f_2)$ can be found.   The method described in section 3.5.2 needed $\lambda_{12}(f_2)$ in functional form in order to get $\lambda_{12}(\underline{x})$ which is used both in a consistency check and in reversion to the decision space. However, in many problems there is no need for a consistency check on λ_{12} since these values can be found accurately by solving minimization problem 3 - 1.   One can then use the approach described in the previous section for reverting to the decision space without needing to know $\lambda_{12}(\underline{x})$.   Thus one can avoid both the regression for $\lambda_{12}(f_2)$ and the problem of transforming that into $\lambda_{12}(\underline{x})$ which for some problems (such as dynamic ones) can give a complicated or unusable result.

However, regressions to find λ_{12} as an analytic function of f_2 can still be employed, if desired by the analyst. For certain problems it may be simpler to use these functions to determine additional values at which to question the DM rather than resolving the ε-constraint problem for each additional value. In general, one would not expect to be able to approximate $\lambda_{12}(f_2)$ by a simple polynomial with any reasonable degree of accuracy but interpolations or curve fitting procedures over small intervals between known values may be useful in determining additional values of λ_{12}.

4.1.5   Finding the Indifference Band

Once the non-inferior solutions have been found, the DM is questioned to find out his assessments of worth. Since the only value of interest is where $W_{12}(f_2) = 0$, the functional form of $W_{12}(f_2)$ need not be found.   One approach to finding the values of f_2 for which $W_{12}(f_2) = 0$ is a type of exhaustive search technique.   In this approach, the DM is asked to assess the worth at   equally spaced non-inferior values of f_2 , e.g. , $f_2^0 ... f_2^k$, until W_{12} changes sign. As soon as W_{12} changes sign, the preferred value of f_2 is known to be between the last two tested values. For example, if $W_{12}(f_2^m)$ has a different sign than $W_{12}(f_2^{m+1})$ then the preferred value of f_2 must be between f_2^m and f_2^{m+1} .   The search procedure can then be restarted with a smaller increment over the interval (f_2^m, f_2^{m+1}). Due to the monotonicity of the surrogate worth function, if two different values of f_2 are preferred solutions, then any value between them will also be a preferred solution. Note that even with this approach, not all of the non-inferior values need be tested.

Another approach is a type of gradient approach or Newton approximation method.   Two values of the worth are found, e. g., $W_{12}(f_2^0)$ and $W_{12}(f_2^1)$.   The next value of f_2 tried is the one where a straight line through the two known values hits $W_{12} = 0$. Mathematically

$$f_2^{K+1} = f_2^K - W_{12}(f_2^K) (f_2^K - f_2^{K-1}) / [W_{12}(f_2^K) - W_{12}(f_2^{K-1})].$$

Once a value of f_2 is found for which $W_{12}(f_2) = 0$, the rest of the indifference band is determined by finding the worth at values near this preferred solution. This latter approach may require fewer questions to the decision maker, but more calculation is needed to determine which question to ask. Also note that for the value of f_2 at which the DM should be questioned $(\hat{f}_2)$, the trade-off ratio may not have been found in the first segment. One could then either resolve the optimization problem 3-1 for $\varepsilon_2 = \hat{f}_2$ to find the trade-off ratio exactly, or use curve-fitting, interpolation or regression techniques on known non-inferior values to approximate the trade-off ratio. Also note that these algorithms are equally applicable when W_{12} is considered as a function of λ_{12}; the same equations hold if f_2 is replaced by λ_{12}.

4.2   THE STATIC TWO-OBJECTIVE ε-CONSTRAINT (STE) ALGORITHM

Algorithms will now be presented describing the computational pro-

cedures for solving two-objective problems with the SWT method.   All of
these algorithms assume that the DM is able to accurately assess his pre-
fererences; they could easily be modified as described in section 4.1.2
if this is an untenable assumption.   The first algorithm uses the ε-con-
straint approach both for finding the non-inferior points and for reverting to the decision space to find the preferred solution.   A flowchart of
this algorithm is provided in figure 4-2.

4.2.1 The Algorithm

Step 1: Find the minimum value, f_{2MIN}, for f_2 by solving:

$$\text{MIN} \quad f_2(\underline{x})$$

$$\text{s.t.} \quad \underline{x} \; \varepsilon \; T$$

The solution to this problem if f_{2MIN}.

Step 2: Find the maximum value, f_{2MAX}, for f_2 by solving:

$$\text{MIN} \quad f_1(\underline{x})$$

$$\text{s.t.} \quad \underline{x} \; \varepsilon \; T$$

If the solution vector to this problem is $\underline{x}^*$ then f_{2MAX} is $f_2(\underline{x}^*)$; if there
is more than one solution vector $\underline{x}^*$ then the minimum value of $f_2(\underline{x}^*)$ is
f_{2MAX} .

Step 3: Set the initial value for $\varepsilon_2 = f_{2MAX} - \Delta$ where $\Delta > 0$.

Step 4: $\text{MIN} \quad f_1(\underline{x})$

$$\text{s.t.} \quad f_2(\underline{x}) \leq \varepsilon_2$$

$$\underline{x} \; \varepsilon \; T$$

Generally the Kuhn-Tucker conditions can be used to solve this problem. Of
course, any optimization technique which is appropriate can be utilized.[2]
Let $\underline{x}^*$ be the decision vector which solves this problem. The solution is
$f_1^*(\varepsilon_2) = f_1(\underline{x}^*)$; each solution should also contain λ_{12}, the Lagrange multiplier for the ε_2 constraint.   Also a check to see if the constraint is
binding must be made as part of this step. If the constraint is binding then
$\varepsilon_2 = f_2$ so that the outputs of this step are $\lambda_{12}(f_2)$ and $f_1^*(f_2)$ at the
value $f_2 = \varepsilon_2$. If the constraint is not binding then ignore these values.

Step 5: If the ε_2 constraint was binding in step 4, then set $\varepsilon_2 =$
$\varepsilon_2 - \Delta$; otherwise set $\varepsilon_2 = f_2(\underline{x}^*)$.

If ε_2 is greater than f_{2MIN} then return to step 3; otherwise   pro-

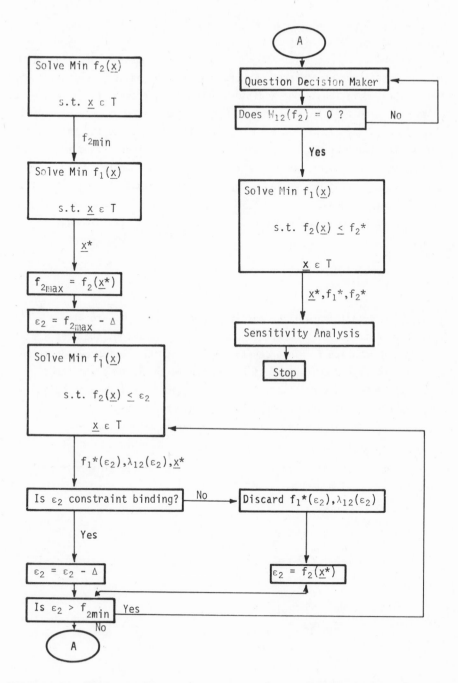

Figure 4-2.   Flowchart for Static Two-Objective ε-Constraint Algorithm.

ceed to step 6.

Step 6: Develop the surrogate worth function $W_{12}(f_2)$ as follows: For each value $\hat{f}_2$ at which the worth is desired, ask the decision maker for his assessment of how much $\lambda_{12}(\hat{f}_2)$ additional units of objective f_1 are worth in relation to one additional unit of objective f_2, given $\hat{f}_2$ units of f_2 and $f_1^*(\hat{f}_2)$ units of f_1. His assessment can be made on a relative scale of say -10 to +10 with zero signifying equal worth; this assessment is the value of W_{12} at $\hat{f}_2$. The search techniques described in section 4.3.5 are used to determine at which values of f_2 the worth should be found.

Step 7. The preceeding step is repeated until the entire indifference band is found.

Step 8. Find the preferred decision vector $\underline{x}^*$ by solving:

$$\text{MIN } f_1(\underline{x})$$
$$\text{s.t. } f_2(\underline{x}) \leqslant f_2^*$$
$$\underline{x} \in T$$

In many problems there will be more than one solution f_2^* to step 7. In that case, there will be more than one preferred solution, and step 8 must be repeated for each f_2^* in order to find all of the preferred decisions vectors; some other criteria must be introduced to decide among them.

Step 9: A sensitivity analysis could be performed to determine the possible effects of implementing the preferred solution.

Step 10: Stop - the solution is ready to be implemented.

4.2.2 Sample Problem

The use of this algorithm will be illustrated by solving example 1 from chapter 1. The problem is:

$$\text{MIN } x_1$$
$$\text{MIN } 10 - x_1 - x_2$$
$$\text{s.t. } 0 \leqslant x_1 \leqslant 5$$
$$0 \leqslant x_2 \leqslant 5$$

Step 1: We find f_{2MIN} by solving:

$$\text{MIN } 10 - x_1 - x_2$$
$$\text{s.t. } 0 \leqslant x_1 \leqslant 5$$
$$0 \leqslant x_2 \leqslant 5$$

The solution to this problem is at $x_1 = 0$, $0 \leqslant x_2 \leqslant 5$. Since it is not unique, take the one which gives the minimum value of $f_2(\underline{x})$ which is $x_1 = 0$, $x_2 = 5$ with $f_2(\underline{x}) = 5$. Thus $f_{2MAX} = 5$.

Step 3: Letting $\Delta = .5$, set $\varepsilon_2 = 4.5$.

Step 4: Solve MIN x_1

$$s.t. \quad 10 - x_1 - x_2 \leqslant \varepsilon_2$$
$$0 \leqslant x_1 \leqslant 5$$
$$0 \leqslant x_2 \leqslant 5$$

The simplex method can be used to solve this for all of the values of ε_2. Reformulating these equations in simplex format

$$MIN \quad x_1$$
$$s.t. \quad x_1 + x_2 - S_1 + R_1 = 10 - \varepsilon_2$$
$$x_1 + S_2 = 5$$
$$x_2 + S_3 = 5$$

where S_1, S_2, S_3 are slack variables and R_1 is the artificial variable. For $\varepsilon_2 = 4.5$ the final tableau is

	z	x_1	x_2	S_1	S_2	S_3	b
z	1	0	0	-1	0	-1	.5
x_1	0	1	0	-1	0	-1	.5
S_2	0	0	0	1	1	1	4.5
x_2	0	0	1	0	0	1	5.0

Thus the solution is .5 so $f_1^*(4.5) = .5$; since S_1 and R_1 both equal zero, the ε-constraint is binding and this value is acceptable. The value of the Lagrange multiplier is the negative of the coefficient under the slack variable corresponding to the ε_2 constraint; thus $\lambda_{12}(4.5) = 1.0$.

Step 5: Since the ε_2 constraint is binding, set $\varepsilon_2 = 4.0$ and go back to step 3. This is repeated until $\varepsilon_2 = 0$ is reached; the results are summarized in table 4-1.

Step 6: The decision maker is questioned and assume his responses are as given in table 4-2.

The exhaustive search type algorithm was used to determine the values of f_2 at which the worth was found. Linear interpolation was used for those non-inferior values not found in table 4-1.

Step 7: The indifference band is found to be $2.0 \leqslant f_2^* \leqslant 2.3$.

TABLE 4 - 1

Results of Static Two-Objective ε-Constraint Problem (Step 5)

f_2	$f_1^*(f_2)$	$\lambda_{12}(f_2)$
4.5	0.5	1.0
4.0	1.0	1.0
3.5	1.5	1.0
3.0	2.0	1.0
2.5	2.5	1.0
2.0	3.0	1.0
1.5	3.5	1.0
1.0	4.0	1.0
0.5	4.5	1.0

TABLE 4 - 2

DM Responses for Static Two-objective ε-Constraint Problem

f_2	$W_{12}(f_2)$
4.5	+ 10
4.0	+ 8
3.5	+ 7.5
3.0	+ 4
2.5	+ 1
2.0	0
2.4	+ 0.5
2.3	0
1.9	- 1

Step 8: Find the preferred decision vector by solving step 4 with ε_2 replaced by each f_2^* in the indifference band. For example, for $f_2^* = 2.0$, the result is $x_1^* = 3.0$,

The preferred decision vector may not be unique. Also note that there may be more than one indifference solution or indifference bands, consequently, the associated preferred decision vectors will likely be different.

Step 9: At this point a sensitivity analysis should be performed but such work is beyond the scope of this text.

Step 10: Stop.

Notice that even though the trade-off function was linear, this algorithm had no problems in finding preferred solutions.

4.3 THE MULTIPLIER APPROACH

One of the major problems with the ε-constraint algorithm is that the minimizations required may be complicated to solve. One way to circumvent this difficulty is to use the multiplier approach, a variation of the parametric method, to find the non-inferior points.

Consider the following problem:

Problem 4-4.

$$\text{MIN } f_1(\underline{x}) + \lambda_{12} \, f_2(\underline{x})$$

$$\text{s.t. } \underline{x} \in T$$

This is equivalent to the parametric approach where $\lambda_{12} = \theta_2/\theta_1$. It can be shown[3] that the solution vector $\underline{x}^*$ to problem 4-4 solves the following problem:

$$\text{MIN } f_1(\underline{x})$$

$$\text{s.t. } f_2(\underline{x}) \leq f_2(\underline{x}^*)$$

where λ_{12} is the Lagrange multiplier for the f_2 constraint. Thus one can set the value of $\lambda_{12} > 0$ and solve problem 4-4, finding the corresponding f_1^* and f_2^* (non-inferior values of the objectives) to be used in determining the worth function. This method avoids the necessity of finding the minimum and maximum values for ε_2 since setting $\lambda_{12} > 0$ assures being in the non-inferior region; also the minimizations are simpler since there is one less constraint.

It is possible for convex two-objective problems to determine a maximum value for λ_{12} in the non-inferior region. Since the trade-off function $f_1^*(f_2)$ is convex, λ_{12} will be a monotonically decreasing function of f_2, so that the largest value of λ_{12} will occur at the minimum value of f_2. Thus the following problem can be solved:

$$\text{MIN } f_1(\underline{x})$$

$$\text{s.t. } f_2(\underline{x}) \le f_{2MIN}$$

$$\underline{x} \in T$$

where f_{2MIN} is found as in section 4.1.1. The Lagrange multiplier corresponding to the f_2 constraint is λ_{12MAX}.

4.3.1 Limitations of the Multiplier Approach

There are several problems that can arise with the multiplier approach. Although all of the solutions found by this approach are non-inferior points, not all of the non-inferior points can be found when the duality gap problem[4] exists. Hopefully, one would be able to generate enough points to accurately determine the worth functions; any given problem would have to be judged on its own merits as to whether or not the information generated by the multiplier approach is adequate.

A more severe problem is that λ_{12} in the multiplier approach does not always correspond to the negative of the slope of the trade-off function. Consider the feasible functional space S shown in figure 4-3. If the slope of the trade-off curve at point A is $- \lambda_{12}^0$ then any line L with slope $-\lambda_{12}$ where $\lambda_{12} \ge \lambda_{12}^0$ will also find the same point A at the minimum. In this case, the values of $\lambda_{12} > \lambda_{12}^0$ do not correspond to $-df_1/df_2$. In terms of duality theory, any specific value λ_{12}^0 may not necessarily correspond to a stationary point of the Lagrangian to problem 3-1.

This occurs whenever there are discontinuities in the slope of the trade-off function and is especially prominent in linear problems. Thus it is advisable to use such an approach only for non-linear problems. In convex non-linear problems, this effect appears only at the end points of the non-inferior set; for non-convex trade-off functions, it can also occur where the duality gaps start and end. Since these inaccuracies are present at only a few points in non-linear problems, the accuracy of the surrogate worth function would not be affected too greatly if enough values were generated.

4.4 THE STATIC TWO-OBJECTIVE COMBINED (STC) ALGORITHM

An algorithm will now be presented using the multiplier approach for the first segment of the solution procedure to lower the computational efforts required, while retaining the ε-constraint approach for the reversion to the decision space. A flowchart for this algorithm is provided in figure 4-4.

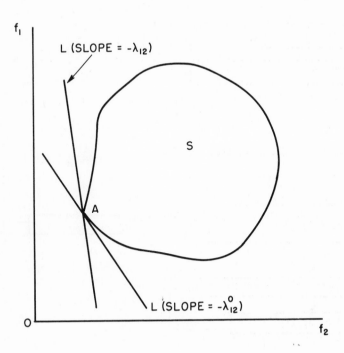

Note: $\lambda_{12} > \lambda_{12}^0$

Both lines find point A at the minimum.

Figure 4-3.   Failure of Multiplier Approach.

4.4.1   The Algorithm

Step 1:   Find the maximum value for λ_{12} by solving:

$$\text{MIN} \quad f_1(\underline{x})$$
$$\text{s.t. } f_2(\underline{x}) \le f_{2MIN}$$
$$\underline{x} \in T$$

where f_{2MIN} is the solution to MIN $f_2(\underline{x})$ s.t. $\underline{x} \in T$. The Lagrange multiplier for the f_2 constraint is λ_{12MAX}. If this step is computationally too difficult one could set $\lambda_{12MAX} = \infty$.

Step 2: Select an initial value $\hat{\lambda}_{12}$ such that $0 < \hat{\lambda}_{12} < \lambda_{12MAX}$.

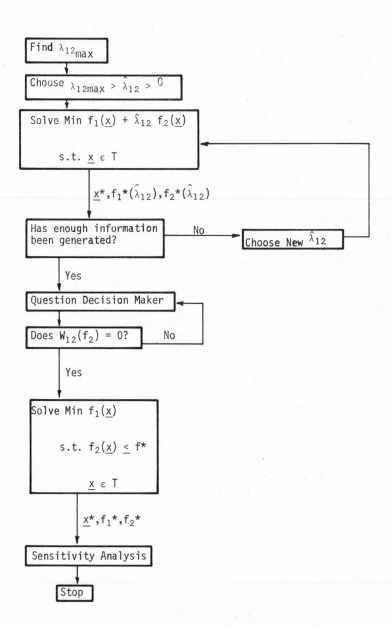

Figure 4-4. Flowchart for Static Two-Objective Combined Algorithm.

Step 3:   Solve MIN $f_1(\underline{x}) + \hat{\lambda}_{12} \; f_2(\underline{x})$

s.t. $\underline{x} \; \varepsilon \; T$

The solution vector $\underline{x}^*$ is substituted into $f_1(\underline{x})$ and $f_2(\underline{x})$ to find $f_1^*(\hat{\lambda}_{12})$ and $f_2^*(\hat{\lambda}_{12})$. These values are not necessarily unique.

Step 4: If enough information has been generated, go on to step 5, if not, choose a new value of $\hat{\lambda}_{12} > 0$ and go back to step 3.

Step 5: Develop the surrogate worth function W_{12} as follows:

For each set of values $\hat{\lambda}_{12}$, $f_1^*(\hat{\lambda}_{12})$, $f_2^*(\hat{\lambda}_{12})$ at which the worth is desired, ask the DM for his assessment of how much λ_{12} additional units of objective f_1 are worth in relation to one additional unit of f_2, given $f_2^*(\hat{\lambda}_{12})$ units of f_2 and $f_1^*(\hat{\lambda}_{12})$ units of f_1. His assessment is then the value of W_{12} at $f_2^*(\hat{\lambda}_{12})$. The search techniques described in section 4.3.5 are used to determine at which values of f_2 the worth should be found.

Step 6: The preceeding step is repeated until the entire indifference band is found.

Step 7: Find the preferred decision vector $\underline{x}^*$ by solving:

$$MIN \;\; f_1(\underline{x})$$
$$s.t. \; f_2(\underline{x}) \leqslant f_2^*$$
$$\underline{x} \; \varepsilon \; T$$

These last two steps are the same as in the ε-constraint approach. Again, if there is more than one solution to step 6, step 7 must be repeated for each of these solutions f_2^* in order to find all of the preferred solutions.

Step 8: A sensitivity analysis could be performed to determine the possible effects of implementing the preferred solution.

Step 9: Stop!

4.4.2 Sample Problem

The use of this algorithm will be illustrated by applying it to the following non-linear problem:

$$MIN \;\; f_1 = x_1^2 + 2 \; x_2^2$$
$$MIN \;\; f_2 = x_1 + 3 \; x_2$$

where x_1 and x_2 are unbounded.

Step 1: Since f_2 is unbounded $(f_{2MIN} = -\infty)$, λ_{12MAX} cannot be determined.

Thus, set $\lambda_{12MAX} = \infty$.

Step 2:   Set $\lambda_{12} = .5$.

Step 3:   Solve MIN f' $= x_1^2 + 2\ x_2^2 + \lambda_{12}(x_1 + 3\ x_2)$.

The necessary condition for a minimum is $\nabla f' = 0$ (it is also sufficient since f' is convex); thus

$$2\ x_1 + \lambda_{12} = 0 \quad \text{or} \quad x_1 = -\ \lambda_{12}/2$$

and $$4\ x_2 + 3\ \lambda_{12} = 0 \quad \text{or} \quad x_2 = -3\ \lambda_{12}/4 \ .$$

Thus for $\lambda_{12} = .5$, $f_1^*(.5) = 11/32$, $f_2^*(.5) = -\ 11/8$.

Step 4:   Different values of λ_{12} are used in step 2,   and   the results obtained are given in table 4-3.

Step 5:   The surrogate worth function is developed, and it is assumed the DM's assessments are as given in table 4-4.

Using the gradient algorithm , it is obvious that the third trial should be made at $f_2 = -\ 2\ 1/8$. The corresponding trade-off is found by linear interpolation from table 4-3 to be $\lambda_{12} = .75$ and $f_1^* = 55/64$. Note that for this problem , a quadratic interpolation would be more accurate for f_1^* . Once $f_2 = -\ 2\ 1/8$ is found to be a preferred solution , other neighboring values are tried to determine the extent of the indifference band.

Step 6:   The indifference band is found to be $-2\ 1/8 \leq f_2^* \leq -\ 2$.

Step 7:   We find $\underline{x}^*$ by solving:

$$\text{MIN} \quad x_1^2 + 2\ x_2^2$$
$$\text{s.t.} \ \ x_1 + 3\ x_2 \leq f_2^*$$

for each f_2^* in the indifference band. Since the constraint must be binding this can be solved by substituting $x_1 = f_2^* - 3\ x_2$ into the objective equation and using $\nabla f_1 = 0$. For example, the preferred solution for $f_2^* = -2.0$ is found to be $x_1^* = -4/11$, $x_2^* = -\ 6/11$.

Step 8:   Sensitivity analysis could be performed here.

Step 9:   Stop!

4.5   THE STATIC TWO-OBJECTIVE MULTIPLIER (STM) ALGORITHM

In order to use the multiplier approach in the second segment also (to revert to the decision space in finding the preferred decision vector

Table 4 - 3

Results of Static Two-Objective Combined Problem (Step 4)

λ_{12}	$f_1^*(\lambda_{12})$	$f_2^*(\lambda_{12})$
0.5	11/32	- 1 3/8
1.0	1 3/8	- 2 3/4
2.0	5 1/2	- 5 1/2
3.0	12 3/8	- 8 1/4
4.0	22	- 11
5.0	35 5/8	- 13 3/4
10.0	137 1/2	- 27 1/2

Table 4 - 4

DM Responses for Static Two-Objective Combined Problem

f_2	$W_{12}(f_2)$
- 1 3/8	+ 1.5
- 2 3/4	- 1.5
- 2 1/8	0
- 2	0
- 2 1/4	- 1

$\underline{x}^*$), it is necessary to know the preferred trade-off rate λ_{12}^*. Since the solution of the surrogate worth function gives the preferred value f_2^* , it would be necessary to know λ_{12} as a function of f_2 in order to find λ_{12}^*. This could be done by performing a regression on the values of f_2 and λ_{12} found in step 3 of the previous algorithm, but this could introduce errors which may be sizable. In some cases, the value of f_2 in the non-inferior region may be known as a function of λ_{12} directly from the necessary conditions used in solving step 3. For these problems one could then find the inverse function $\lambda_{12}(f_2)$; thus $\lambda_{12}^* = \lambda_{12}(f_2^*)$. For non-convex problems, using the multiplier approach to revert to the decision space may cause some of the preferred solutions to be missed due to the possibility of duality gaps. However, if the other algorithms prove impossible to solve, this approach can be used anyway to get one solution. Finding $\lambda_{12}(f_2)$ would replace steps 5 and 6 below.

This algorithm's accuracy is guaranteed only for problems in which the trade-off function is non-linear and convex (where f_2 is a one-to-one function of λ_{12} over the range of f_2 in the non-inferior set). For these cases, the development of the worth function W_{12} as a function of λ_{12} is guaranteed to be valid. The solution of $W_{12}(\lambda_{12}) = 0$ gives the preferred trade-off rate λ_{12}^* which can then be used in a parametric procedure. A flowchart for this algorithm is given in figure 4-5.

4.5.1 The Algorithm

Steps 1 through 4 are the same as in the Static Two-objective Combined algorithm and will not be repeated here.

Step 5: Develop the surrogate worth function $W_{12}(\lambda_{12})$ as follows: For each set of values $\hat{\lambda}_{12}$, $f_1^*(\hat{\lambda}_{12})$, $f_2^*(\hat{\lambda}_{12})$ at which the worth is desired, ask the DM for his assessment of how much $\hat{\lambda}_{12}$ additional units of objective f_1 are worth in relation to one additional unit of objective f_2 given $f_2^*(\hat{\lambda}_{12})$ units of f_2 and $f_1^*(\hat{\lambda}_{12})$ units of f_1. His assessment is then the value of W_{12} at λ_{12}. The search techniques described in section 4.3.5 are used to determine at which values of λ_{12} the worth should be found.

Step 6: The preceeding step is repeated until the entire indifference band is found.

Step 7: Find the preferred decision vector $\underline{x}^*$ by solving:

$$\text{MIN} \quad f_1(\underline{x}) + \lambda_{12}^* \, f_2(\underline{x})$$

$$\text{s.t.} \quad \underline{x} \, \epsilon \, T$$

This is the same problem as step 3 with λ_{12} replaced by λ_{12}^*; thus little additional calculation is needed.

Step 8: A sensitivity analysis could be performed here to determine the possible effects of implementing the preferred solution.

Step 9: Stop!

4.5.2 Sample Problem

Consider the same problem as in section 4.4.2 It can be shown that f_2 is a one to one function of λ_{12} as follows. The necessary conditions developed in the first segment of that example (in step 3) showed that $x_1 = -\lambda_{12}/2$ and $x_2 = 3\lambda_{12}/4$. Substituting back into $f_2(\underline{x})$ gives $f_2(\lambda_{12}) = -11\lambda_{12}/4$ for any value of $\lambda_{12} > 0$. Thus f_2 is a one-to-one function of λ_{12} and the Static Two-objective Multiplier algorithm is appropriate. Since steps 1 through 4 are the same as in section 4.4.2, we continue with

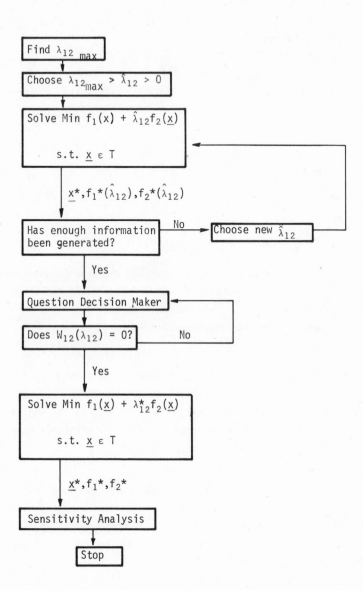

Figure 4-5. Flowchart for Static Two-Objective Multiplier Algorithm.

step 5.

Step 5: The surrogate worth function is developed; assume that the DM's assessments are the same as in section 4.4.2. These are listed as a function of λ_{12} in table 4-5.

Table 4 - 5

DM Responses for Static Two-Objective Multiplier Problem

λ_{12}	$W_{12}(\lambda_{12})$
0.5	+ 1.5
1.0	- 1.5
0.75	0
0.7	0
0.6	+ 1
0.8	- 1

The gradient search type algorithm was used to determine the values of λ_{12} at which the worth was found. Linear interpolation was used for those non-inferior values not found in table 4-3.

Step 6: The indifference band is found to be $.7 \leq \lambda_{12}^{*} \leq .75$.

Step 7: Solve MIN $x_1^2 + 2\, x_2^2 + \lambda_{12}^{*}(x_1 + 3\, x_2)$.

Taking the gradient and setting it equal to zero for $\lambda_{12}^{*} = .75$, preferred values of the decision variables are found to be $x_1^{*} = -3/8$, $x_2^{*} = - 9/16$. This can be repeated for other values of λ_{12}^{*} in the indifference band. Thus approximately the same result is obtained with this algorithm as with the static two-objective combined algorithm.

4.6   SUMMARY

This chapter has presented some modification in computational method from the version of the surrogate worth trade-off method described in chapter 3 to make it more generally applicable and easier to use. Three algorithms for implementing the SWT method for the static two-objective case were presented along with discussions of their applicability and simple examples of their use. The ε-constraint (STE) algorithm is applicable to all problems but may be difficult computationally to solve. The combined (STC) algorithm can be applied to non-linear problems with a probable savings in the computation and only a slight loss of accuracy. The multiplier (STM) algorithm is guaranteed only when the trade-off function is convex and non-linear, but is generally the most efficient procedure.

Of course , one should keep in mind that it is difficult to state which algorithm is best for any given problem ; it is up to the analyst to use his discretion in determining which one is appropriate for his specific situation.

The question of how well the mathematical equations in the objective functions represent society's true objectives is a common problem in modeling. For example, to what extent does the number of man-hours of usage adequately measure recreational objectives for a reservoir? These problems are not explicitly taken into account in the SWT method; however, they can be somewhat ameliorated during the process of interacting with the decision maker by insuring that he understands exactly what he is deciding on. In general , one should exercise caution when using the SWT method with objectives for which measures are either poorly defined or mathematical equations unavailable.

As in all approaches there will probably exist certain pathological problems in which none of the algorithms are applicable. For example, an inability to solve the Kuhn-Tucker conditions could lead to the inability to find non-inferior points. Thus, one must be cautious when applying these algorithms to perverse functions. The next chapter will apply the SWT method to dynamic problems.

FOOTNOTES

1. Note that the " * " is used to denote both non-inferior values and preferred values; it should be clear from the contex which is intended.

2. See Wagner [1969], Hillier and Lieberman [1967], Intriligator [1971], Taha [1971], or any text on optimization techniques.

3. This was first proven by Everett [1963].

4. Again see Everett [1963], Lasdon [1968], or Gembicki [1973].

REFERENCES

1. Everett, H.III, "Generalized Lagrange Multiplier Method for Solving Problems of Optimum Allocation of Resources," Operations Research vol. 11, 1963.

2. Gembicki, F., "Vector Optimization for Control with Performance and Parameter Sensitivity Indices," Ph.D. Dissertation, Case Western Reserve University, 1963.

3. Hillier, F.S. and Lieberman, G.J., Introduction to Operations Re-

search, Holden-Day, San Francisco, 1967.

4.   Intriligator, M.D., Mathematical Optimization and Economic Theory,
     Prentice-Hall Inc., Englewood Cliffs, N.J., 1971.

5.   Lasdon, L.S.,"Duality and Decomposition in Mathematical Programming"
     IEEE Transactions, vol. SSC-4, no. 2, 1968.

6.   Taha, H.A., Operations Research; An Introduction, The Macmillan
     Company, N.Y., 1971.

7.   Wagner, H.M., Principles of Operations Research with Applications
     to Managerial Decisions, Prentice Hall Inc., Englewood Cliffs,
     N. J., 1969.

Chapter 5

THE SWT METHOD FOR DYNAMIC TWO-OBJECTIVE PROBLEMS

The preceeding chapters have attempted to solve problems where the objectives are static functions of a vector of decision variables. This chapter will consider the application of the SWT method to dynamic systems where the objectives are functions of the state of the system as well as of the decision (control) variables, with both states and decisions time dependent. It is assumed that the state of the system at any time is a known function of time and the previous states and decisions. The following notation will be used: $C^n[0,t_f]$ is the set of all continuous functions from the closed interval $[0,t_f]$ into R^n; $\dot{x}(t)$ is the first derivative of x with respect to the independent variable t .

It will be assumed that the ith objective can be formulated as follows:

$$f_i = \phi_i(\underline{x}(t_f)) + \int_0^{t_f} a_i(\underline{x}(t),\underline{u}(t),t) \, dt, \quad i = 1,2, \ldots, n$$

where $\underline{x}(t)\varepsilon\ C^p[0,t_f]$ is a vector of state variables, $\underline{u}(t)\varepsilon\ C^r[0,t_f]$ is a vector of control variables, and t_f is the terminal value of the independent variable which will be considered fixed and the same for all objectives. The integral term can be viewed as summing the contributions (a_i) to the objective all along the trajectory, while ϕ_i is the contribution to the objective of the final state of the system.

Note, however, that not all dynamic problems can be put into this form with f_i scalar valued. For example, the values of a_i at different t may be non-commensurable. Multiplying factors such as the interest rate are commonly used to commensurate the values of a_i at different t . In general, however, a_i at each t would be a different non-commensurable objective and the problem would become an infinite-objective problem.

There are various types of constraints which may arise in dynamic problems. End point constraints of the form $\underline{g}(\underline{x}(t_f),t_f) \leqslant \underline{0}$ will be included in this formulation. Path constraints, such as $N(\underline{x}(t),\underline{u}(t),t) \leqslant 0$ can be also included in the following development[1] , but problems with constraints of this form tend to be rather difficult to actually solve; for the sake of clarity, they will be avoided in this formulation. Also note that problems where t_f is a control variable (e.g. minimum time problems) can be handled by modifying the necessary conditions for a minimum.

This chapter will show the relationship between the static and dynamic problems. The dynamic problem will be put into ε-constraint form and the Lagrange multipliers for these constraints will be shown to again represent the elements of the trade-off rate matrix. Algorithms that are analogous to the static case will be presented for the solution of two objective dynamic problems, including sample problems to illustrate their use.

5.1   INTRODUCTORY ANALYSIS

The general multiple objective dynamic problem can be written in vector notation as:

$$\text{MIN } \underline{f} = \underline{\phi}(\underline{x}(t_f)) + \int_0^{t_f} \underline{a}(\underline{x}(t),\underline{u}(t),t) \, dt$$

$$\text{s.t. } \underline{\dot{x}}(t) = \underline{\psi}(\underline{x}(t),\underline{u}(t),t) \; ; \; \underline{x}(0) \text{ given}$$

$$\underline{g}(\underline{x}(t_f), t_f) \leq \underline{0}$$

where $\underline{f}$, $\underline{a}$, and $\underline{\phi}$ are the n-vectors whose elements are f_i, a_i, and ϕ_i, respectively.

The feasible decision space T will be a subset of $C^r[0,t_f]$ and will thus be complicated (or impossible) to work with. However, one can again look at the feasible function space $S = \{f | \underline{\dot{x}}(t) = \underline{\psi}(\underline{x}(t),\underline{u}(t),t)$ with $\underline{x}(0)$ as given and $\underline{g}(\underline{x}(t_f),t_f) \leq \underline{0}\}$. Note that S is a subset of R^n. Since the surrogate worth trade-off method operates primarily in the functional space S, this method will be easily adaptable to dynamic problems. The non-inferior set will again be on the boundary of S, and can be represented by the trade-off function, with its slope represented by the trade-off rate functions. These will also be scalar valued and independent of time and can therefore be determined and utilized just as in the static case. Thus the only difference between the static and dynamic problems occurs in the decision space; the functional space, S, is identical in both cases.

5.2   DYNAMIC PROBLEMS IN ε-CONSTRAINT FORM

The two objective dynamic problem can be put into ε-constraint form directly as follows:

$$\text{MIN } \phi_1(\underline{x}(t_f)) + \int_0^{t_f} a_1(\underline{x}(t),\underline{u}(t),t)dt$$

$$\text{s.t.}\quad \dot{\underline{x}}(t) = \underline{\Psi}(\underline{x}(t),\underline{u}(t),t);\ \underline{x}(0) \text{ given}$$

$$\underline{g}(\underline{x}(t_f),t_f) \leq \underline{0}$$

$$\phi_2(\underline{x}(t_f)) + \int_0^{t_f} a_2(\underline{x}(t),\underline{u}(t),t)\, dt \leq \varepsilon_2$$

However, since the constraint in this form is not very useful, the following substitution is made.

Define a new state variable y such that $\dot{y}(t) = a_2(\underline{x}(t),\underline{u}(t),t)$ and $y(0) = 0$.

$$\text{Then}\quad y(t_f) = \int_0^{t_f} a_2(\underline{x}(t),\underline{u}(t),t)dt,$$

$$\text{and}\quad f_2 = y(t_f) + \phi_2(\underline{x}(t_f)) .$$

The problem then becomes:

$$\text{MIN}\quad \phi_1(\underline{x}(t_f)) + \int_0^{t_f} a_1(\underline{x}(t),\underline{u}(t),t)dt$$

$$\begin{aligned} \text{s.t.}\quad & \dot{\underline{x}}(t) = \underline{\Psi}(\underline{x}(t),\underline{u}(t),t);\ \underline{x}(0) \text{ given} \\ & \dot{y}(t) = a_2(\underline{x}(t),\underline{u}(t),t);\quad y(0) = 0 \\ & \underline{g}(\underline{x}(t_f),t_f) \leq \underline{0} \\ & y(t_f) + \phi_2(\underline{x}(t_f)) \leq \varepsilon_2 \end{aligned}$$

Thus the constraint is now included as an end point constraint and can be solved by any of the available methods.[2]

It will now be shown that the Lagrange multiplier for the ε_2 constraint (when it is binding) is the value of the trade-off rate function at $f_2 = \varepsilon_2$. First form the Lagrangian:

$$\begin{aligned} L = {} & \phi_1(\underline{x}(t_f)) + \lambda[y(t_f) + \phi_2(\underline{x}(t_f)) - \varepsilon_2] + \underline{\mu}^T[\underline{g}(\underline{x}(t_f),t_f)] \\ & + \int_0^{t_f} \{a_1(\underline{x}(t),\underline{u}(t),t) + \underline{\nu}_1^T(t)[\underline{\Psi}(\underline{x}(t),\underline{u}(t),t) - \dot{\underline{x}}(t)] \\ & + \nu_2(t)[a_2(\underline{x}(t),\underline{u}(t),t) - \dot{y}(t)]\}dt \end{aligned}$$

where λ is a scalar Lagrange multiplier for the ε_2 constraint, $\underline{\mu}$ is an m-vector of multipliers for the end point constraints, $\underline{\nu}_1(t)$ is a p-vector of Lagrange multipliers (which are functions of time) for the system equation constraints, and $\nu_2(t)$ is a scalar Lagrange multiplier (also a function of time) for the system equation constraints for the new state variable y . It can now be seen that $\partial L/\partial \epsilon_2 = -\lambda$.

As in the static case, only those values of ϵ_2 which correspond to points in the non-inferior region will be considered. Thus, the ϵ_2 constraint must be binding. Again, when the optimum is found, $L = \phi_1(\underline{x}(t_f)) + \int_0^{t_f} a_1(\underline{x}(t),\underline{u}(t),t)\ dt = f_1^*$. Since the ϵ-constraint is binding, $\epsilon_2 = y(t_f) + \phi_2(\underline{x}(t_f)) = f_2$. Thus $\lambda = -\partial f_1^*/\partial f_2$ and the Lagrange multiplier for the ϵ_2 constraint is really $\lambda_{12}(\epsilon_2)$, i.e., the trade-off rate function evaluated at $f_2 = \epsilon_2$.

5.3   DYNAMIC TWO-OBJECTIVE ϵ-CONSTRAINT (DTE) ALGORITHM

This section will present algorithms describing the computational procedures for solving multiple objective dynamic problems which are analogous to the static algorithms. The first of these uses the ϵ-constraint approach for both finding the non-inferior points and for reverting to the decision space to find the preferred decision vector. Note that all of the computational efficiencies developed in the functional space in chapter 4 will be applicable to dynamic problems. In particular the use of W_{12} as a function of f_2, the methods for finding limits on f_2, the avoidance of regressions and the assumption of a consistent DM will be utilized. A flowchart for this algorithm is presented in figure 5-1.

5.3.1   The Algorithm

Step 1: Find the minimum value for f_2 by solving:

$$\text{MIN}\quad f_2 = \phi_2(\underline{x}(t_f)) + \int_0^{t_f} a_2(\underline{x}(t),\underline{u}(t),t)dt$$

$$\text{s.t. } \underline{\dot{x}}(t) = \underline{\psi}(\underline{x}(t),\underline{u}(t),t); \underline{x}(0)\quad \text{given}$$

$$\underline{g}(\underline{x}(t_f),t_f) \leqslant \underline{0}$$

The solution to this problem is f_{2MIN} .

Step 2: Find the maximum value for f_2 by solving:

$$\text{MIN}\quad f_1 = \phi_1(\underline{x}(t_f)) + \int_0^{t_f} a_1(\underline{x}(t),\underline{u}(t),t)dt$$

$$\text{s.t. } \underline{\dot{x}}(t) = \underline{\psi}(\underline{x}(t),\underline{u}(t),t); \underline{x}(0)\quad \text{given}$$

$$\text{Solve Min } \left\{ \phi_2(\underline{x}(t_f)) + \int_0^{t_f} a_2(\underline{x}(t),\underline{U}(t),t) \; dt \right\}$$

$$\text{s.t. } \underline{\dot{x}}(t) = \underline{\psi}(\underline{x}(t),\underline{u}(t),t) \; ; \; \underline{x}(0) \text{ given}$$

$$\underline{g}(\underline{x}(t_f),t_f) \leq \underline{0}$$

f_{2min}

$$\text{Solve Min } \left\{ \phi_1(\underline{x}(t_f)) + \int_0^{t_f} a_1(\underline{x}(t),\underline{u}(t),t) \; dt \right\}$$

$$\text{s.t. } \underline{\dot{x}}(t) = \underline{\psi}(\underline{x}(t),\underline{u}(t),t) \; ; \; \underline{x}(0) \text{ given}$$

$$\underline{g}(\underline{x}(t_f),t_f) \leq \underline{0}$$

$\underline{x}^*(t),\underline{u}^*(t)$

$$f_{2max} = \phi_2(\underline{x}^*(t_f)) + \int_0^{t_f} a_2(\underline{x}^*(t),\underline{u}^*(t),t) \; dt$$

$$\varepsilon_2 = f_{2max} - \Delta$$

$$\text{Solve Min } \left\{ \phi_1(\underline{x}(t_f)) + \int_0^{t_f} a_1(\underline{x}(t),\underline{u}(t),t) \; dt \right\}$$

$$\text{s.t. } \underline{\dot{x}}(t) = \underline{\psi}(\underline{x}(t),\underline{u}(t),t) \; ; \; \underline{x}(0) \text{ given}$$

$$\dot{y}(t) = a_2(\underline{x}(t),\underline{u}(t),t) \; ; \; y(0) = 0$$

$$\underline{g}(\underline{x}(t_f),t_f) \leq \underline{0}$$

$$y(t_f) + \phi_2(\underline{x}(t_f)) \leq \varepsilon_2$$

A   B

Figure 5-1. Flowchart for Dynamic Two-Objective ε-Constraint Algorithm.
Continued next page.

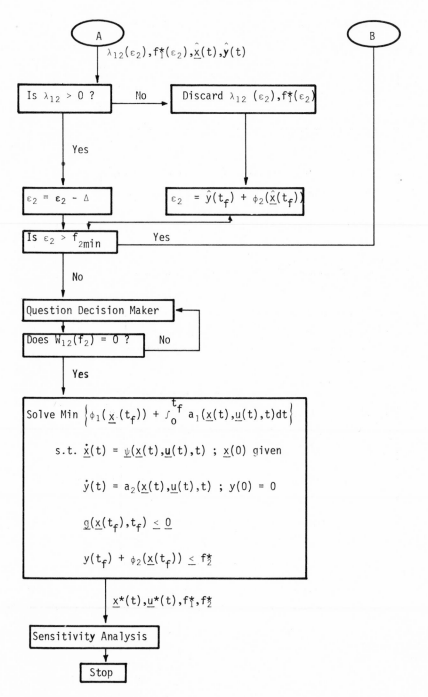

Figure 5-1.   Continued

$$\underline{g}(\underline{x}(t_f), t_f) \leq \underline{0}$$

If the state vector $\underline{x}^*(t)$ and control vector $\underline{u}^*(t)$ solve this problem,

then $f_{2MAX} = \phi_2(\underline{x}^*(t_f)) + \int_0^{t_f} a_2(\underline{x}^*(t), \underline{u}^*(t), t)dt$. If there is more than

one state vector $\underline{x}^*(t)$ and control vector $\underline{u}^*(t)$ solving this problem, then

f_{2MAX} is the minimum value of $\phi_2(\underline{x}^*(t_f)) + \int_0^{t_f} a_2(\underline{x}^*(t), \underline{u}^*(t), t)dt$.

Step 3: Set the initial value $\varepsilon_2 = f_{2MAX} - \Delta$ where $\Delta > 0$.

Step 4: Solve the problem

$$\text{MIN} \quad \phi_1(\underline{x}(t_f)) + \int_0^{t_f} a_1(\underline{x}(t), \underline{u}(t), t)dt$$

$$\text{s.t.} \quad \underline{\dot{x}}(t) = \underline{\psi}(\underline{x}(t), \underline{u}(t), t); \quad \underline{x}(0) \quad \text{given}$$

$$\dot{y}(t) = a_2(\underline{x}(t), \underline{u}(t), t); \quad y(0) = 0$$

$$\underline{g}(\underline{x}(t_f) t_f) \leq \underline{0}$$

$$y(t_f) + \phi_2(\underline{x}(t_f)) \leq \varepsilon_2$$

The general approach to solving this problem is to define the Hamiltonian:

$$H(\underline{x}(t), \underline{u}(t), t) = a_1(\underline{x}(t), \underline{u}(t), t) + \underline{v}_1^T(t) \cdot \underline{\psi}(\underline{x}(t), \underline{u}(t), t) + v_2(t) \ .$$

The necessary conditions for a minimum then become (the arguments of the functions are omitted for clarity):

1) $\quad \partial H / \partial \underline{u} = 0$

2a) $\quad \underline{\dot{v}}_1^T = - \partial H / \partial \underline{x}$

2b) $\quad \underline{v}_1^T(t_f) = \dfrac{\partial \phi_1}{\partial \underline{x}} + \lambda_{12} \dfrac{\partial(y + \phi_2)}{\partial \underline{x}} + \underline{\mu}^T \dfrac{\partial \underline{g}}{\partial \underline{x}} \bigg|_{t = t_f}$

$\qquad\qquad = \dfrac{\partial \phi_1}{\partial \underline{x}} + \lambda_{12} \dfrac{\partial \phi_2}{\partial \underline{x}} + \underline{\mu}^T \dfrac{\partial \underline{g}}{\partial \underline{x}} \bigg|_{t = t_f}$

3a) $\quad \dot{v}_2 = - \partial H / \partial y$

3b) $\quad v_2(t_f) = \dfrac{\partial \phi_1}{\partial y} + \lambda_{12} \dfrac{\partial(y + \phi_2)}{\partial y} + \underline{\mu}^T \dfrac{\partial \underline{g}}{\partial y} \bigg|_{t = t_f} = \lambda_{12}$

4) $\quad \underline{\dot{x}} = \underline{\psi} \quad ; \quad \underline{x}(0) \quad \text{given}$

5) $\dot{y} = a_2$; $y(0) = 0$

6) $\underline{\mu}^T[\underline{g}] = 0$; $\underline{\mu} \geq \underline{0}$

7) $\lambda_{12}[(y+\phi_2) - \varepsilon_2]_{t=t_f} = 0$; $\lambda_{12} \geq 0$

The first six conditions are solved as in any optimal control problem to find $\underline{\hat{u}}(t), \underline{\hat{x}}(t)$ and $\hat{y}(t)$ in terms of λ_{12}. Condition 7 can then be used to find $\hat{\lambda}_{12}$. Assume that the ε_2 constraint is binding, or $y(t_f)+\phi_2(\underline{x}(t_f))= \varepsilon_2$. Then substituting $\hat{y}(t_f)$ and $\underline{\hat{x}}(t_f)$ found from the first six conditions, this equation can be solved for $\hat{\lambda}_{12}$. If the value $\hat{\lambda}_{12}$ is less than zero, then the second part of the condition $(\lambda_{12} \geq 0)$ is violated and the assumption of a binding constraint was therefore incorrect; thus $\hat{\lambda}_{12} = 0$ and the constraint is non-binding. Since a non-binding constraint means the solution is in the inferior region, this solution will be ignored. If $\hat{\lambda}_{12}$ is greater than zero, then the assumption was correct and the solution is in the non-inferior region. $\hat{\lambda}_{12}$ is then the value of $\lambda_{12}(f_2)$ at $f_2 = \varepsilon_2$. The value $f_1^*(f_2)$, which corresponds to the solution of the minimization problem, can be found by substituting $\hat{\lambda}_{12}$ back into $\underline{\hat{x}}(t)$ and $\underline{\hat{u}}(t)$.

$$\text{Then}\quad f_1^*(f_2) = \phi_1(\underline{\hat{x}}(t_f)) + \int_0^{t_f} a_1(\underline{\hat{x}}(t),\underline{\hat{u}}(t),t)dt.$$

Step 5: If the ε_2 constraint was binding in step 4, then $\varepsilon_2 = \varepsilon_2 - \Delta$. Otherwise lower ε_2 to the value attained by f_2 at the solution to step 4 - set $\varepsilon_2 = \hat{y}(t_f) + \phi_2(\underline{\hat{x}}(t_f))$, where $\hat{y}(t)$ and $\underline{\hat{x}}(t)$ are the values found in step 4 with the value $\hat{\lambda}_{12} = 0$ substituted. Note that changing ε_2 only changes condition 7 of the necessary conditions, so little extra work is required to obtain more non-inferior points. If ε_2 is greater than f_{2MIN} return to step 3; otherwise continue on to step 6.

Step 6: Develop the surrogate worth function $W_{12}(f_2)$ exactly as in the static case; recall that this can be achieved since f_2 is scalar valued. For each value $\hat{f}_2$ at which the worth is desired, ask the DM for his assessment of how much $\lambda_{12}(\hat{f}_2)$ additional units of objective f_1 are worth in relation to one additional unit of objective f_2, given $\hat{f}_2$ units of f_2 and $f_1^*(\hat{f}_2)$ units of f_1. His assessment is made on an ordinal scale, say from -10 to +10 with zero signifying equivalent worth; this assessment is the value of W_{12} at f_2. The search techniques described in section 4.3.5 are used to determine at which values of f_2 the worth should be found.

Step 7: The preceeding step is repeated until the entire indifference band is found.

Step 8: Find the preferred state vectors $\underline{x}^*(t)$ and control vectors $\underline{u}^*(t)$ by solving step 4 with ε_2 replaced by f_2^* for each f_2^* in the indifference band:

$$\text{MIN} \quad \phi_1(\underline{x}(t_f)) + \int_0^{t_f} a_1(\underline{x}(t),\underline{u}(t),t)dt$$

$$\text{s.t.} \quad \underline{\dot{x}}(t) = \underline{\psi}(\underline{x}(t),\underline{u}(t),t) \quad ; \quad x(0) \quad \text{given}$$

$$\dot{y}(t) = a_2(\underline{x}(t),\underline{u}(t),t) \quad ; \quad y(0) = 0$$

$$\underline{g}(\underline{x}(t_f),t_f) \leq 0$$

$$y(t_f) + \phi_2(\underline{x}(t_f)) \leq f_2^*$$

Since the necessary conditions 1 through 6 are the same as in step 4, they do not have to be resolved. Also the constraint must be binding since f_2^* must correspond to a non-inferior point. Thus, little extra computation is necessary to revert to the decision space. If there is more than one f_2^* in step 7, then there is more than one preferred solution, and step 8 must be repeated for each f_2^* in order to find all of the preferred solutions; some other criteria must then be introduced to decide among them.

Step 9: A sensitivity analysis could be performed to determine the possible effects of implementing the preferred solution.

Step 10: Stop!

In most real problems, it will be impossible to analytically solve the simultaneous boundary value differential equations in the necessary conditions of steps 1, 2, 4 and 8. However, there are numerical approximation techniques available which can often be used, e.g., quasilinearization, gradient methods, and neighboring extremal methods.[3]

5.3.2   Sample Problem

The use of this algorithm will be illustrated with the following example:

Example 5-1: $\text{MIN} \quad x(1) + \int_0^1 u^2(t) \, dt$

$$\text{MIN} \quad x(1) + \int_0^1 (u(t) - 5)^2 \, dt$$

$$\text{s.t.} \quad \underline{\dot{x}}(t) = u(t) \quad ; \quad x(0) = 10$$

where x and u are $\varepsilon\ C^1\ [0,1]$.

Step 1: Find f_{2MIN} by solving:

$$MIN\ x(1) + \int_0^1 (u - 5)^2\ dt$$

$$s.t.\ \dot{x} = u\ ;\ x(0) = 10$$

The Hamiltonian for this problem is $H = (u - 5)^2 + \nu\ u$

The necessary conditions are:

1) $\dfrac{\partial H}{\partial u} = 2\ (u - 5) + \nu = 0$

2) $\dot{\nu} = - \dfrac{\partial H}{\partial x} = 0\ ;\ \nu(1) = 1$

3) $\dot{x} = u\ ;\ x(0) = 10$

The second condition implies $\nu(t) = 1$ which can be substituted into condition 1 giving $u(t) = 9/2$. Since $u(t)$ is a constant, condition 3 can be easily integrated to give $x(t) = 10 + 9\ t/2$. Thus the solution to this problem is $f_{2MIN} = 14.75$.

Step 2: Find f_{2MAX} by solving:

$$MIN\ x(1) + \int_0^1 u^2\ dt$$

$$s.t.\ \dot{x} = u\ ;\ x(0) = 10$$

The Hamiltonian for this problem is $H = u^2 + \nu\ u$.
The necessary conditions are:

1) $\dfrac{\partial H}{\partial u} = 2\ u + \nu = 0$

2) $\dot{\nu} = - \dfrac{\partial H}{\partial x} = 0\ ;\ \nu(1) = 1$

3) $\dot{x} = u\ ;\ x(0) = 10$

The second condition implies $\nu(t) = 1$ which can be substituted into condition 1 giving $u(t) = - 1/2$. Condition 3 is then integrated to give $x(t) = 10 - t/2$. These values of $x(t)$ and $u(t)$ are substituted into $f_2 = x(1) +$

$\int_0^1 (u - 5)^2\, dt$ to give f_{2MAX} = 39.75.

Step 3: Set the initial value for ε_2 = 39.0; for other iterations Δ will be 1.0.

Step 4: Reformulate the problem in ε-constraint form:

$$\text{MIN} \quad x(1) + \int_0^1 u^2\, dt$$

$$\text{s.t.} \quad \dot{x} = u \; ; \quad x(0) = 10$$

$$\dot{y} = (u - 5)^2 \; ; \quad y(0) = 0$$

$$x(1) + y(1) \le \varepsilon_2$$

The Hamiltonian for this problem is $H = u^2 + \nu_1 u + \nu_2 (u - 5)^2$.

The necessary conditions are:

1) $\dfrac{\partial H}{\partial u} = 2u + \nu_1 + 2\nu_2 (u-5) = 0$

2) $\dot{\nu}_1 = -\dfrac{\partial H}{\partial x} = 0 \; ; \quad \nu_1(1) = 1 + \lambda_{12}$

3) $\dot{\nu}_2 = 0 \; ; \quad \nu_2(1) = \lambda_{12}$

4) $\dot{x} = u \; ; \quad x(0) = 10$

5) $\dot{y} = (u - 5)^2 \; ; \quad y(0) = 0$

6) $\lambda_{12}[y(1) + x(1) - \varepsilon_2] = 0 \; ; \quad \lambda_{12} \ge 0$

Conditions 2 and 3 give $\nu_1(t) = 1 + \lambda_{12}$ and $\nu_2(t) = \lambda_{12}$.

Substitution into condition 1 gives:
$$u(t) = (9 \lambda_{12} - 1)/(2 + 2 \lambda_{12}) \tag{1}$$

Integrating conditions 4 and 5 gives:
$$x(t) = 10 + (9 \lambda_{12} - 1) \, t \, /(2 + 2 \lambda_{12}) \tag{2}$$
$$y(t) = (11 + \lambda_{12})^2 \, t \, /(2 + 2 \lambda_{12})^2 \tag{3}$$

Assume that the ε_2 constraint is binding so that:
$$y(1) + x(1) = \varepsilon_2 \tag{4}$$

Substituting equations (2) and (3) into (4) gives:
$$(59 - 4 \varepsilon_2) \lambda_{12}^2 + (118 - 8 \varepsilon_2) \lambda_{12} + (159 - 4 \varepsilon_2) = 0 \tag{5}$$

For ε_2 = 39.0 the solution is λ_{12} = .0153; the negative root for λ_{12} is ig-nored because this would not meet the $\lambda_{12} \geq 0$ requirement. $f_1^*(39.0)$ is now found by substituting λ_{12} = .0153 into the expressions for $x(t)$ and $u(t)$ (equations (1) and (2)) and solving .

$$f_1 = x(1) + \int_0^1 u^2\, dt; \quad \text{the result is } f_1^*(39.0) = 9.7557 \; .$$

Step 5: Since the constraint was binding, set ε_2 = 38.0 and go back to step 3. This is repeated until ε_2 = 14.0; the results are summarized in table 5-1. As a check λ_{21} was found at values of f_1 corresponding to the values in column 2 of table 5-1; the error between λ_{12} and $1/\lambda_{12}$ was less than .0001 in all cases.

Step 6: The decision maker is questioned and it is assumed that his responses are as given in table 5-2.

The exhaustive search type algorithm was used to determine the values of f_2 at which the worth was found. Linear interpolation was used for those non-inferior values not found in table 5-1.

Table 5 - 1

Results of Dynamic Two-Objective ε-Constraint Problem (Step 5)

f_2	$f_1^*(f_2)$	$\lambda_{12}(f_2)$
15.0	30.00	9.00
16.0	24.82	3.47
17.0	22.00	2.33
18.0	19.97	1.77
19.0	18.38	1.43
20.0	17.09	1.18
21.0	16.00	1.00
22.0	15.07	0.86
23.0	14.28	0.74
24.0	13.59	0.64
25.0	12.98	0.56
26.0	12.46	0.49
27.0	12.00	0.43
28.0	11.60	0.37
29.0	11.25	0.32
30.0	10.95	0.28

Table 5-1 (Cont'd)

f_2	$f_1^*(f_2)$	$\lambda_{12}(f_2)$
31.0	10.69	0.24
32.0	10.47	0.20
33.0	10.28	0.17
34.0	10.13	0.14
35.0 *	10.00	0.11
36.0	9.90	0.08
37.0	9.83	0.06
38.0	9.78	0.04
39.0	9.76	0.02

Table 5 - 2

DM Responses for Dynamic Two-Objective ε-Constraint Problem

f_2	$W_{12}(f_2)$
15.0	- 7
16.0	- 6
17.0	- 5
18.0	- 4
19.0	- 2
20.0	- 0.5
21.0	0
22.0	+ 0.5
20.5	- 0.5
21.5	+ 0.5

Step 7: One preferred value is $f_2^* = 21.0$ as can be seen directly from table 5-2. Note that the indifference band is very small for this decision maker.

Step 8: The value $f_2^* = 21.0$ can be used directly in place of ε_2 in equation (5) to get the preferred trade-off rate $\lambda_{12}^* = 1.0$; this is then substituted into equations (1) and (2) to get the preferred state variable $x^*(t) = 10 + 2t$ and preferred control variable $u^*(t) = 2$. The preferred value of objective f_1 can also be found to be 16.0 .

5.4   DYNAMIC TWO-OBJECTIVE COMBINED (DTC) ALGORITHM

One of the major problems with the ε-constraint algorithm is that the minimizations required are often quite complicated to solve.   To lower the computational efforts required, an algorithm is developed using the multiplier approach for the first segment of the solution procedure;   the ε-constraint approach will still be used for the reversion to the decision space.   Just as in the static case one can define a new objective:

$$f' = f_1 + \lambda_{12}f_2 = \phi_1(\underline{x}(t_f)) + \lambda_{12}\phi_2(\underline{x}(t_f)) + \int_0^{t_f} \{a_1(\underline{x}(t),\underline{u}(t),t)$$

$$+ \lambda_{12} \; a_2(\underline{x}(t),\underline{u}(t),t)\} \quad dt$$

If f' is minimized subject to the same system constraints, $\dot{\underline{x}}(t) = \underline{\psi}(\underline{x}(t),\underline{u}(t),t),\underline{x}(0)$ given, and $\underline{g}(\underline{x}(t_f),t_f) \leq \underline{0}$, then this is equivalent to the parametric approach with $\lambda_{12} = \theta_2/\theta_1$. Since setting $\lambda_{12} > 0$ guarantees a non-inferior point, the steps of finding limits for ε_2 can be avoided.   Maximum values for λ_{12} can be found, as in the static case, by putting the problem in ε-constraint form and solving it with ε_2 replaced by f_{2MIN} where f_{2MIN} is found as in the previous algorithm; the multiplier corresponding to the ε_2 constraint is then λ_{12MAX}.   However, it is felt that the need for λ_{12MAX} is not large enough to justify the greatly increased computational burden.   If λ_{12MAX} is not known, one could keep increasing λ_{12} until two consecutive values give the same non-inferior point.

Since it involves one less constraint, this method should be simpler to solve than using the ε-constraint approach. The limitations of this approach are the same as for the static case; it is subject to inaccuracies for problems where the trade-off function is non-convex or linear.   By the same reasoning as in the static case (see section 4.3.1), the use of this method is not recommended for linear problems.   However, it may be used in other cases with hopefully small inaccuracies. A flowchart of this algorithm is provided in figure 5-2.

5.4.1 The Algorithm

Step 1:   Set $\lambda_{12} = \hat{\lambda}_{12} > 0$

Step 2:   Solve the following problem:

$$MIN \; \phi_1(\underline{x}(t_f)) + \hat{\lambda}_{12}\phi_2(\underline{x}(t_f)) + \int_0^{t_f} \{a_1(\underline{x}(t),\underline{u}(t),t))$$

$$+ \hat{\lambda}_{12}a_2(\underline{x}(t),\underline{u}(t),t)\} \; dt$$

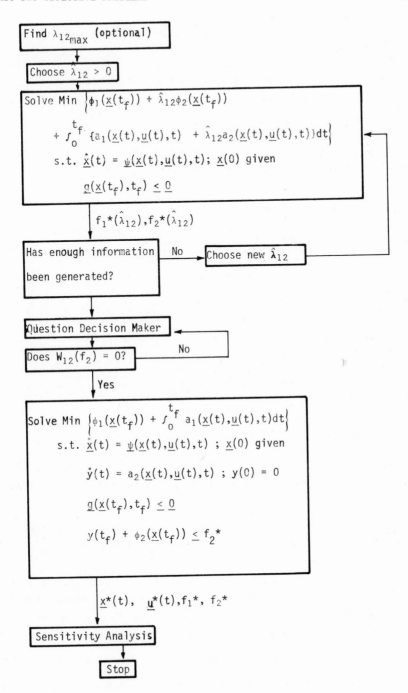

Figure 5-2. Flowchart for Dynamic Two-objective Combined Algorithm.

$$\text{s.t.} \quad \dot{\underline{x}}(t) = \underline{\psi}(\underline{x}(t), \underline{u}(t), t) \quad ; \quad \underline{x}(0) \quad \text{given}$$

$$\underline{g}(\underline{x}(t_f), t_f) \leq \underline{0}$$

The general approach to solving this problem is to define the Hamiltonian:

$$H(\underline{x}(t), \underline{u}(t), t) = a_1(\underline{x}(t), \underline{u}(t), t) + \hat{\lambda}_{12} a_2(\underline{x}(t), \underline{u}(t), t)$$

$$+ \underline{\nu}_1^T \cdot \underline{\psi}(\underline{x}(t), \underline{u}(t), t)$$

Again the arguments of the functions will be dropped for simplicity. The necessary conditions for a minimum are:

1) $\quad \dfrac{\partial H}{\partial \underline{u}} = 0$

2) $\quad \dot{\underline{\nu}}_1 = -\dfrac{\partial H}{\partial \underline{x}} \; ; \; \underline{\nu}_1(t_f) = \dfrac{\partial \phi_1}{\partial \underline{x}} + \hat{\lambda}_{12} \dfrac{\partial \phi_2}{\partial \underline{x}} + \underline{\mu}^T \dfrac{\partial \underline{g}}{\partial \underline{x}} \Big|_{t=t_f}$

3) $\quad \dot{\underline{x}} = \underline{\psi} \; ; \; \underline{x}(0) \quad \text{given}$

4) $\quad \underline{\mu}^T \underline{g} = 0 \; ; \; \underline{u} \geq \underline{0}$

These are solved as in any optimal control problem to find $\hat{\underline{x}}(t)$ and $\hat{\underline{u}}(t)$ which can then be substituted into the original objectives to find

$$f_1^*(\hat{\lambda}_{12}) = \phi_1(\hat{\underline{x}}(t_f)) + \int_0^{t_f} a_1(\hat{\underline{x}}(t), \hat{\underline{u}}(t), t) \, dt \quad \text{and} \quad f_2^*(\hat{\lambda}_{12}) = \phi_2(\hat{\underline{x}}(t_f))$$

$$+ \int_0^{t_f} a_2(\hat{\underline{x}}(t), \hat{\underline{u}}(t), t) \, dt$$

For any value of $\hat{\lambda}_{12}$ there may be more than one solution $\hat{\underline{x}}(t)$ and $\hat{\underline{u}}(t)$; in that case there is more than one f_1^* and f_2^* corresponding to the trade-off rate value $\hat{\lambda}_{12}$.

Step 3:   If it is felt that enough information has already been generated, then proceed to step 4. If not, chose a new value of $\hat{\lambda}_{12} > 0$ and return to step 2.

Step 4:   Develop the surrogate worth function $W_{12}(f_2)$. For each set of values $\hat{\lambda}_{12}$, $f_1^*(\hat{\lambda}_{12})$, $f_2^*(\hat{\lambda}_{12})$ at which the worth is desired, ask the DM for his assessment of how much $\hat{\lambda}_{12}$ additional units of objective f_1 are worth in relation to one additional unit of f_2, given $f_2^*(\hat{\lambda}_{12})$ units of f_2 and $f_1^*(\hat{\lambda}_{12})$ units of f_1. His assessment is then the value of W_{12} at $f_2^*(\hat{\lambda}_{12})$. The search techniques described in section 4.3.5 are used to determine at which values of f_2 the worth should be found.

Step 5:   The preceeding step is repeated until the entire indifference band is found.

Step 6:   Find the preferred state vectors $\underline{x}^*(t)$ and control vectors $\underline{u}^*(t)$ by solving the following problem for each f_2^* in the indifference band

$$\text{MIN}\quad \phi_1(\underline{x}(t_f)) + \int_0^{t_f} a_1(\underline{x}(t),\underline{u}(t),t)\ dt$$

$$\text{s.t. } \underline{\dot{x}}(t) = \underline{\Psi}(\underline{x}(t),\underline{u}(t),t)\ ;\ \underline{x}(0)\ \text{given}$$

$$\dot{y}(t) = a_2(\underline{x}(t),\underline{u}(t),t)\ ;\ y(0) = 0$$

$$\underline{g}(\underline{x}(t_f),t_f) \leq \underline{0}$$

$$y(t_f) + \phi_2(\underline{x}(t_f)) \leq f_2^*$$

The constraint will be binding since f_2^* must be in the non-inferior set, so the inequality constraint can be replaced by an equality. This can be solved just as in the ε-constraint method by application of the necessary conditions.   If there is more than one f_2^* in step 5, this step must be repeated for each one in order to determine all of the preferred solutions. This reversion to the decision space does add a great deal of computational complexity, but the entire algorithm is still simpler computationally than the ε-constraint algorithm.

Step 7:   A sensitivity analysis can be performed to determine the possible effects of implementing the preferred solution.

Step 8:   Stop!

5.4.2 Sample Problem

The use of this algorithm will be illustrated with the same problem as was used for the Dynamic Two Objective ε-constraint algorithm in section 5.3.2.

Example 5-2: $\text{MIN}\ f_1 = x(1) + \int_0^1 u^2(t)\ dt$

$$\text{MIN}\ f_2 = x(1) + \int_0^1 (u(t) - 5)^2\ dt$$

$$\text{s.t. } \dot{x}(t) = u(t)\ ;\ x(0) = 10$$

$$x,u \in C^1[0,1]$$

Step 1:   Set $\lambda_{12} = 1.0$

Step 2:   Solve   MIN $(1 + \lambda_{12})$ $x(1)$ + $\int_0^1 u^2 + \lambda_{12}(u-5)^2$ dt

s.t. $\dot{x}$ = u ; x(0) = 10

The Hamiltonian for this problem is H = $u^2 + \lambda_{12}(u-5)^2 + \nu_1 u$.
The necessary conditions for a minimum are:

1) $\frac{\partial H}{\partial u}$ = $2u + 2\lambda_{12}(u-5) + \nu_1$ = 0

2) $\dot{\nu}_1$ = $- \frac{\partial H}{\partial x}$ = 0 ; $\nu_1(1)$ = 1 + λ_{12}

3) $\dot{x}$ = u ; x(0) = 10

Condition 2 implies $\nu_1(t)$ = 1 + λ_{12}; substituting into condition 1 gives
u (t) = $(9\lambda_{12} - 1)/(2 + 2\lambda_{12})$.   Integrating condition 3 yields:
x(t) = 10 + $(9\lambda_{12} -1)t/(2+2\lambda_{12})$.
Substituting into f_1 and f_2 yields:

$f_1^*(\lambda_{12})$ = $(9\lambda_{12}-1)^2/(2+2\lambda_{12})^2$ + 10 + $(9\lambda_{12} - 1)/(2+2\lambda_{12})$

$f_2^*(\lambda_{12})$ = $(11 + \lambda_{12})^2/(2+2\lambda_{12})^2$ + 10 + $(9\lambda_{12}-1)/(2+2\lambda_{12})$

For λ_{12} = 1.0, $f_1^*(1.0)$ = 16.0, $f_2^*(1.0)$ = 21.0 .

Step 3:   The problem is solved for other values of λ_{12} > 0 and the
results are summarized in table 5-3.

Step 4:   The DM is questioned - assume his responses are as given
in table 5-4.

Step 5: One preferred value is f_2^* = 21.0 as can be seen directly.
The rest of the indifference band could be found by questioning the DM at
value of f_2 near 21.0.

Step 6:   Solve the problem:

MIN x(1) + $\int_0^1 u^2$ dt

s.t. $\dot{x}$ = u ; x(0) = 10

$\dot{y}$ = $(u-5)^2$; y(0) = 0

y(1) + x(1) $\leq$ 21.0

The Hamiltonian for this problem is H = $u^2 + \nu_1 u + \nu_2(u-5)^2$.
The necessary conditions are:

TABLE 5 - 3

Results of Dynamic Two-Objective Combined Problem (Step 3)

λ_{12}	$f_1^*(\lambda_{12})$	$f_2^*(\lambda_{12})$
.25	10.75	30.75
.5	12.53	25.86
.75	14.34	22.91
1.0	16.00	21.00
2.5	22.50	16.79
5.0	27.11	15.44
7.5	29.21	15.10
10.0	30.41	14.96

TABLE 5 - 4

DM Responses for Dynamic Two-Objective Combined Problem

f_2	$W_{12}(f_2)$
14.96	- 10
15.10	- 6.5
15.44	- 6
16.79	- 5.5
21.00	0
22.91	+ 1

1) $\dfrac{\partial H}{\partial u} = 2u + \nu_1 + 2\nu_2(u-5) = 0$

2) $\dot{\nu}_1 = \dfrac{\partial H}{\partial x} = 0 \ ; \ \nu_1(1) = 1 + \lambda_{12}$

3) $\dot{\nu}_2 = 0 \ ; \ \nu_2(1) = \lambda_{12}$

4) $\dot{x} = u \ ; \ x(0) = 10$

5) $\dot{y} = (u-5)^2 \ ; \ y(0) = 0$

6) $y(1) + x(1) = 21.0$

These equations can be solved as in the example in section 5.3.2. The solution is found to be $x^*(t) = 10 + 2t$, $u^*(t) = 2$, $f_1^* = 16.0$, $f_2^* = 21.0$.

This solution method was much simpler than the dynamic two-objective ε-constraint method.

The latter required the solution of a quadratic equation for each non-inferior point found, while this approach required only a direct substitution. The difference in computational complexity is even more pronounced for more complicated problems. Since in this problem the trade-off function is convex and non-linear, there is no loss of accuracy with this method. The reversion to the decision space requires more effort with the dynamic two-objective combined algorithm, but it does not offset the gain in the first segment.

5.5   DYNAMIC TWO-OBJECTIVE MULTIPLIER (DTM) ALGORITHM

It is obvious that to use the multiplier approach for reversion to the decision space would make the computation even easier. In order to accomplish this, the preferred trade-off rate λ_{12}^* corresponding to the preferred value f_2^* must be known. In some problems the non-inferior value of f_2 may be known as an analytic function of λ_{12} from the necessary conditions in step 2 of the dynamic two-objective combined algorithm. One could then find the inverse function $\lambda_{12}(f_2)$ and thus know $\lambda_{12}^* = \lambda_{12}(f_2^*)$. In other cases, a regression could be performed on the values of λ_{12} and f_2 found in step 2, although this may introduce large errors. For non-convex problems, using the multiplier method for reverting to the decision space may also cause some of the preferred solutions to be missed due to the possibility of duality gaps. However, if the other methods prove to be impossible to solve, this approach may be used anyway, replacing steps 4 and 5 below. Thus, this algorithm's accuracy is guaranteed only for problems with non-linear, convex trade-off functions(when f_2 is a one-to-one function of λ_{12} over the range of f_2 in the non-inferior set). For these cases, the best way of finding λ_{12}^* is to use the approach of developing the worth W_{12} as a function of λ_{12}. The solution of $W_{12}(\lambda_{12}) = 0$ gives the preferred trade-off rate λ_{12}^*. A flowchart of this algorithm is given in figure 5-3.

5.5.1   The Algorithm

Steps 1 through 3 are the same as in the dynamic two-objective combined algorithm and will not be repeated here.

Step 4: Develop the surrogate worth function $W_{12}(\lambda_{12})$. For each set of values $\hat{\lambda}_{12}$, $f_1^*(\hat{\lambda}_{12})$, $f_2^*(\hat{\lambda}_{12})$ at which the worth is desired, ask the DM for his assessment of how much $\hat{\lambda}_{12}$ additional units of objective f_1 are worth in relation to one additional unit of f_2 given $f_2^*(\hat{\lambda}_{12})$ units of f_2 and $f_1^*(\hat{\lambda}_{12})$ units of f_1. His assessment is then the value of W_{12} at $\hat{\lambda}_{12}$. The search techniques described in section 4.3.5 are used to determine at

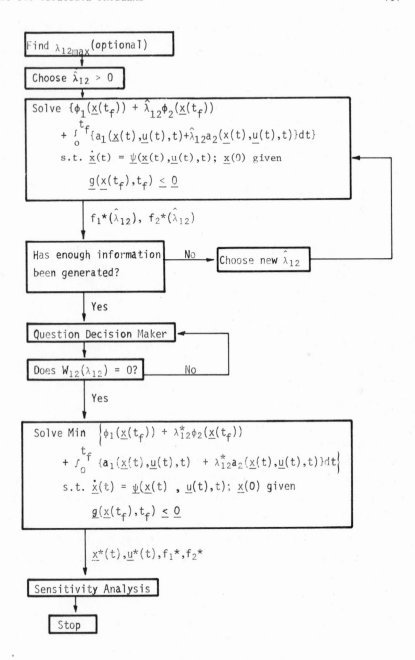

Figure 5-3. Flowchart for Dynamic Two-Objective Multiplier Algorithm.

which values of λ_{12} the worth should be found.

Step 5: The preceeding step is repeated until the entire indifference band is found.

Step 6:   Find the preferred state vectors $\underline{x}^*(t)$ and control vectors $\underline{u}^*(t)$ by solving the following problem for each λ_{12}^* in the indifference band.

$$\text{MIN}\quad \phi_1(\underline{x}(t_f)) + \lambda_{12}^*\, \phi_2(\underline{x}(t_f)) + \int_0^{t_f} \{a_1(\underline{x}(t),\underline{u}(t),t)$$

$$+ \lambda_{12}^*\, a_2(\underline{x}(t),\underline{u}(t),t)\}\ dt$$

$$\text{s.t.}\ \underline{\dot{x}}(t) = \underline{\Psi}(\underline{x}(t),\underline{u}(t),t)\ ;\ \underline{x}(0)\ \text{given}$$

$$\underline{g}(\underline{x}(t_f),t_f) \leqslant \underline{0}$$

Note that this is identical to step 2 with $\hat{\lambda}_{12}$ replaced by λ_{12}^*; thus the necessary conditions need not be resolved and little extra computation is necessary.

Step 7: A sensitivity analysis could be performed to determine the possible effects of implementing the preferred solution.

Step 8: Stop!

This procedure is by far the easiest to solve.

5.5.2 Sample Problem

This algorithm will be illustrated with the same problem as the previous two algorithms.   Since steps 1 through 3 are the same as in example 5-2 (section 5.4.2) they will not be repeated here.

Step 4: The surrogate worth function $W_{12}(\lambda_{12})$ is developed and it is assumed that the DM's assessments are the same as given in example 5-2. These are given as a function of λ_{12} in table 5-5.

Step 5: One preferred value λ_{12}^* is 1.0 as can be seen directly. The rest of the indifference band could be found by questioning the DM at values of λ_{12} near 1.0.

Step 6: Solve the problem:

$$\text{MIN}\ 2\ x(1) + \int_0^1 \{u^2 + (u-5)^2\}\ dt$$

$$\text{s.t.}\quad \dot{x} = u\ ;\ x(0) = 10$$

Since the necessary conditions are the same as in step 2, they can be used to give $x^*(t) = 10 + 2t$, $u^*(t) = 2$, $f_2^* = 21.0$, $f_1^* = 16.0$. It is obvious

TABLE 5 - 5

DM Responses for Dynamic Two-Objective Multiplier Problem

λ_{12}	$W_{12}(\lambda_{12})$
10.0	- 10
7.5	- 6.5
5.0	- 6
2.5	- 5.5
1.0	0
.75	+ 1

from this simple example that the dynamic two-objective multiplier algorithm
provides a great saving in computational effort.

5.6   SUMMARY

This chapter has demonstrated the application of the surrogate worth
trade-off method to dynamic problems.   Three algorithms for implementing
the SWT method for the dynamic two objective case were presented, along
with discussions of their applicability and simple examples of their use.
The ε-constraint (DTE) algorithm is applicable to all problems but may be
difficult to solve   (this difficulty is even more salient in dynamic problems than in static ones).   The combined (DTC) algorithm can be applied to
non-linear problems with a great saving in computation for the first segment (finding non-inferior points) but the reversion to the decision space
which uses the ε-constraint technique may still present difficulties;   in
addition, the use of the multiplier technique in the first segment may introduce inaccuracies.   The multiplier algorithm (DTM) is guaranteed only
when the trade-off function is convex and non-linear, but effects a great
savings in computational effort. Thus, if the ε-constraint algorithm proves
difficult or impossible to solve, the multiplier algorithm may be used,
despite its inaccuracies, to at least get an approximate answer. The next
chapter will consider the use of the SWT method in problems with more than
two objectives.

FOOTNOTES

1. The modifications necessary for such constraints can be found in any optimal control text such as Bryson and Ho [1969], or Athans and Falb [1966].

2. Again see Bryson and Ho [1969] or Athans and Falb [1966].

3. Ibid.

REFERENCES

1. Athans, M. and Falb, P. L., Optimal Control; An Introduction to the Theory and Its Applications, McGraw-Hill, N. Y., 1966.

2. Bryson, A. E., and Ho, Y. C., Applied Optimal Control, Ginn and Co., Waltham, Mass., 1969.

Chapter 6

THE SWT METHOD FOR STATIC n-OBJECTIVE PROBLEMS

The previous chapters have considered the application of the SWT method to the special class of multiple objective problems with only two objectives. These approaches will be generalized in this and the next chapter to problems with n objectives. The definition of the surrogate worth function W_{ij} and preferred solutions will be modified, and the assumption of an accurate decision maker will be studied. The computational efficiencies found for two objective problems will be extended to the n objective case, and algorithms for the solution of these problems will be presented.

6.1 SURROGATE WORTH FUNCTIONS

Consider the n-objective problem in ε-constraint form:

Problem 6-1:

$$\text{MIN } f_1(\underline{x})$$

$$\text{s.t. } \underline{f}(\underline{x}) \leq \underline{\varepsilon}$$

$$\underline{x} \in T$$

where $\underline{f} \in R^{n-1}$ is the vector[1] $(f_2, f_3, \ldots, f_n)^T$ and $\underline{\varepsilon} \in R^{n-1}$ is the vector $(\varepsilon_2, \varepsilon_3, \ldots, \varepsilon_n)^T$. If all the constraints are binding, then the solution to this problem produces the values of $\underline{\Lambda}_1$ and f_1^* at $\underline{f} = \underline{\varepsilon}$, where $\underline{\Lambda}_1 \in R^{n-1}$ is the vector of Lagrange multipliers $(\lambda_{12}, \lambda_{13}, \ldots, \lambda_{1n})^T$ and f_1^* is the non-inferior value of f_1 at fixed values of the other objectives. It is obvious that for any j, λ_{1j} and f_1^* will depend on all of the values f_k, k = 2,3,...,n and not just f_j; thus each λ_{1j} is a function of $\underline{f} = (f_2, f_3, \ldots, f_n)^T$, and the non-inferior value of f_1, f_1^*, is also a function of $\underline{f}$.

At this point, a review of the definition of the surrogate worth function is appropriate.

Definition 6-1: The value of the surrogate worth function W_{ij} is the decision maker's assessment of how much (say on a scale from - 10 to + 10 with zero signifying indifference) he prefers trading λ_{ij} marginal units of f_i for one marginal unit of f_j, given the values of all of the objectives $f_1, \ldots, f_n$ corresponding to λ_{ij}. Note that $W_{ij} > 0$ means that the DM does prefer making such a trade, $W_{ij} < 0$ means that the DM prefers not to make such a trade, and $W_{ij} = 0$ implies indifference.

Since the remainder of this book deals only with finding the surrogate worth, the word "worth" will also be used to mean surrogate worth.

In order to assess the worth W_{1j} at a given value $\hat{f}_j$ for the jth objective, the DM will need to know the values λ_{1j} and f_1^* corresponding to $\hat{f}_j$. However, for any given $\hat{f}_j$ there will be many different values of λ_{1j} and f_1^*, depending on the values of f_k, $k = 2,3,\ldots,n$, $k \neq j$, and thus there will be many different values of the worth. Therefore, the worth cannot be considered as a function of f_j alone; W_{1j} must also be a function of $\underline{f}$. The value of W_{1j} at $\underline{f}$ is the DM's assessment of how much $\lambda_{1j}(\underline{f})$ additional units of f_1 are worth in relation to one additional unit of f_j, given the $i-1^{st}$ component of $\underline{f}$ units of objective f_i (for $i = 2,3,\ldots,n$) and $f_1^*(\underline{f})$ units of f_1. One could also extend the approach of section 3.5.2 of considering W_{1j} as a function of λ_{1j}; in this extension W_{1j} will be a function of $\underline{\Lambda}_1$. Just as in the two objective case, this can only be done when $\underline{\Lambda}_1$ is a one-to-one function of $\underline{f}$ or else the worth at some values of $\underline{\Lambda}_1$ will not be unique. It is possible as a first approximation to consider W_{ij} as a function of f_j alone in order to narrow down the range of the preferred solution.

6.2   PREFERRED SOLUTIONS AND CONSISTENCY

Any element λ_{ij} of the trade-off rate matrix (the matrix of all λ_{ij} for i, $j = 1,2, \ldots,n, i \neq j$) can be considered a function of $\underline{f}$; the first segment of the SWT method would produce λ_{ij} as a function of $(f_1,f_2,\ldots,f_{i-1}, f_{i+1},\ldots,f_n)$ but since f_1 is a function of $f_2,f_3,\ldots,f_n$ in the non-inferior region, λ_{ij} can be considered a function of $(f_2,f_3,\ldots,f_n) = \underline{f}$. Since $\lambda_{ij} = -\partial f_i/\partial f_j\big|_{\underline{f}}$, Euler's chain rule for partial derivatives $(\frac{\partial x}{\partial y} = -\frac{\partial z}{\partial y} / \frac{\partial z}{\partial x})$ can be used to get the following relationship between the λ_{ij} :

$$\lambda_{ij}(\underline{f}) = \lambda_{ik}(\underline{f})\ \lambda_{kj}(\underline{f}) \tag{1}$$

Similarly, using the fact that $\partial x/\partial y = 1/\frac{\partial y}{\partial x}$, it follows that:

$$\lambda_{ij}(\underline{f}) = 1/\lambda_{ji}(\underline{f}) \tag{2}$$

The generalization of the previous section shows that W_{ij} is a function of $(f_1,f_2,\ldots,f_{i-1}, f_{i+1}, \ldots,f_n)$. However, since f_1 is a function of $f_2,f_3, \ldots,f_n$, W_{ij} can be considered a function of $\underline{f}$. When there are n objectives there are $n^2 - n$ worth functions $W_{ij}(\underline{f})$, $i = 1,2, \ldots,n$, $j = 1,2, \ldots,n$, $i \neq j$. Each value $W_{ij}(\underline{f})$ indicates the DM's assessment of "how far" $\lambda_{ij}(\underline{f})$ is from the negative of the rate of change (in the direction of

f_j) of the social indifference surface, at the point in the functional space $(f_1^*(\underline{f});\underline{f})$. This is impossible to depict graphically, but can be viewed as the generalization of figure 3-3.

The actual numerical value of the worth functions are again only relative assessments, but when all of the worth functions simultaneously equal zero, the social indifference and trade-off functions are tangent. Thus the following definitions can be made.

Definition 6-2: A preferred solution is defined as any non-inferior point $(f_1^*;\underline{f}^*)$ in the functional space such that all $W_{ij}(\underline{f}^*) = 0$ for $i = 1,2,...n$, $j = 1,2,...n$, $j \neq i$.

Definition 6-3: A preferred decision vector is defined as any feasible decision vector $\underline{x}^*$ such that $f_1(\underline{x}^*) = f_1^*$ and $\underline{f}(\underline{x}^*) = \underline{f}^*$.

The relationships between the worth functions will now be derived. There are two assumptions that must be made. The first is that the trade-off rate λ_{ij} is a good approximation to the change which would occur in the non-inferior value of f_i when f_j is changed by one unit and all the other objectives are held constant. In general this will only be exactly true in the limit as the increments go to zero, but will be true as an approximation as long as $|f_j| \gg 1$ and $|f_i| \gg |\lambda_{ij}|$. If this assumption is not valid, then the questions asked of the DM should be modified so that he is comparing $c\lambda_{ij}$ units of f_i with c units of f_j, where $c > o$ is a number such that $c\lambda_{ij}$ is a good approximation to the change which would occur in the non-inferior value of f_i when f_j is changed by c units and all the other objectives are held constant. Note that if c is too small, then $c\lambda_{ij}$ will be indistinguishable from zero to the DM, and he will be unable to answer the questions.

The second assumption is that the decision maker is able to accurately assess his preferences in the sense of always identifying the same values for the indifference band. An experiment was performed by Feinberg[2] to determine how well decision makers could assess their preferences. He found that in a three objective problem (power, cost and economy in automobile selection), decision makers would trade power for cost, cost for economy and economy for power at some points. This means that the DM is, in reality, indifferent at these points and yet is unable to recognize this fact. That is, he is assigning positive or negative values to the worth when he has no basis for it. A conclusion is that in practice, one must be careful in excluding from the indifference band values whose worth functions are near zero.

If the DM is unable to accurately assess his preferences, then there will generally be no point at which all of the $W_{ij}(\underline{f}) = 0$. Therefore, it will be assumed that the DM is reasonably accurate. An alternative is discussed at the end of this section.

Theorem 6-1:   If $W_{ij}(\underline{f}) = 0$ then $W_{ji}(\underline{f}) = 0$.

Proof: $W_{ij}(\underline{f}) = 0$ means that there is an equal preference for $\lambda_{ij}(\underline{f})$ marginal units of f_i and one marginal unit of f_j at the non-inferior point $(f_1^*(\underline{f});\underline{f})$. Note that the worth of trading any constant multiple c $\lambda_{ij}(\underline{f})$ marginal units of f_i for c marginal units of f_j at $(f_1^*(\underline{f});\underline{f})$ is the same for any value of c for which the first assumption holds. Specifically, let $c = 1/\lambda_{ij}(\underline{f})$; then there is an equal preference for one marginal unit of f_i and $1/\lambda_{ij}(\underline{f})$ marginal units of f_j. Using equation (2) it is found that there is an equal preference for one marginal unit of f_i and $\lambda_{ji}(\underline{f})$ marginal units of f_j at $(f_1^*(\underline{f});\underline{f})$. Then by the definition of $W_{ji}(\underline{f})$, it follows that $W_{ji}(\underline{f}) = 0$.   Q.E.D.

Unfortunately, there is no simple relationship between the values $W_{ij}(\underline{f})$, $W_{kj}(\underline{f})$ and $W_{ij}(\underline{f})$ because of the fact that the scales on which the worth is assessed are ordinal and thus two different values of the worth cannot be added or multiplied with any meaning. However, it can still be shown that n-1 of the worth functions $(W_{ij}(\underline{f})$, $j = 1,2, \ldots n, j \neq i)$ are sufficient to determine the preferred solutions. The following relationships will be used.

Theorem 6-2: Given $\underline{f}^*$ such that $W_{ij}(\underline{f}^*) = 0$ and $W_{ik}(\underline{f}^*) = 0$ where $j \neq k$, then $W_{jk}(\underline{f}^*) = 0$.

Proof: $W_{ij}(\underline{f}^*) = 0$ implies that $W_{ji}(\underline{f}^*) = 0$ from theorem 6-1, so that there is an equal preference for $\lambda_{ji}(\underline{f}^*)$ marginal units of f_j and one marginal unit of f_i at $(f_1^*(\underline{f}^*);\underline{f}^*)$; similarly $W_{ik}(\underline{f}^*) = 0$ implies $W_{ki}(\underline{f}^*) = 0$ so that there is an equal preference for $\lambda_{ki}(\underline{f}^*)$ marginal units of f_k and one marginal unit of f_i at $(f_1^*(\underline{f}^*);\underline{f}^*)$. A consistent DM yields an equal preference for $\lambda_{ji}(\underline{f}*)$ marginal units of f_j and $\lambda_{ki}(\underline{f}*)$ marginal units of f_k. Thus, $\lambda_{ji}(\underline{f}*)/\lambda_{ki}(\underline{f}*)$ marginal units of f_j and one marginal unit of f_k will be equally preferred. Using equation (1) shows that λ_{jk} marginal units of f_j and one marginal unit of f_k are equally preferred at the point $(f_1^*(\underline{f}^*);\underline{f}^*)$. Thus $W_{jk}(\underline{f}^*) = 0$.   Q.E.D.

At this point the definition of preferred solution can be modified.

<u>Definition 6-4</u>: A preferred solution is defined as any non-inferior point $(f_1^*; \underline{f}^*)$ in the functional space such that $W_{1j}(\underline{f}^*) = 0$ for $j = 2,3, \ldots, n$. Note that the difference between definitions 6-2 and 6-4 is that in the former the property $W_{ij}(\underline{f}^*) = 0$ must hold for all i, while in the latter it need hold only for $i = 1$.

This new definition will be shown to be equivalent to the previous one by the following theorem:

<u>Theorem 6-3</u>: Solving $W_{1j}(\underline{f}) = 0$ simultaneously for $j = 2,3,\ldots$, n is equivalent to solving $W_{ij}(\underline{f}) = 0$ for $i = 1,2, \ldots,n$, $j = 1,2,\ldots,n$, $j \neq i$. Proof: Obviously if $\underline{f}^*$ solves $W_{ij}(\underline{f}) = 0$ for all i and j, then it solves the subset $W_{1j}(\underline{f}) = 0$ for $j = 2,3, \ldots$, n. If $\underline{f}^*$ solves $W_{1j}(\underline{f}) = 0$ for $j = 2,3,\ldots,n$, then by theorem 6-2, $W_{ij}(\underline{f}^*) = 0$ for any $i = 1,2,\ldots,n$, $j = 2,3,\ldots,n$, $i \neq j$, since $W_{1i}(\underline{f}^*) = 0$ and $W_{1j}(\underline{f}^*) = 0$. Also $W_{i1}(\underline{f}^*) = 0$ for any $i = 2,3,\ldots,n$ from the theorem 6-1. Therefore, all of the $W_{ij}(\underline{f}) = 0$ for $i = 1,2,\ldots,n, j = 1,2,\ldots,n$, $i \neq j$ and the two problems are equivalent. Q.E.D.

If the DM is unable to accurately assess his preference then there will generally be no point at which all of the $W_{ij}(\underline{f}) = 0$. In this case, one could solve the sets of n-1 simultaneous equations $W_{ij}(\underline{f}) = 0$; $j = 1,2,\ldots,n$, $j \neq i$ separately for each i to get n different solutions $\underline{f}_1^*$ and then define the preferred solution $\underline{f}^*$ as the average:

$$\underline{f}^* = \sum_{i=1}^{n} \underline{f}_1^*/n.$$ However, for the remainder of this book it will be assumed that the inaccuracies of the DM are negligible.

6.3 COMPUTATIONAL EFFICIENCIES

This section will attempt to extend the computational efficiencies developed for the two objective case to n objective problems. Limits on the values for ε_j, reversion to the decision space to find the preferred decision vector, and the use of regressions and search techniques will be studied.

6.3.1 <u>Limits on ε_j</u>

Unfortunately, when there are more than two objectives, it is impossible to use the same approach as in the two-objective case to determine a maximum value in the non-inferior region for the objectives f_j, $j=2,3,\ldots,n$. Recall that when there were only two objectives, the maximum value of f_2 was found by finding the solution $\underline{x}^*$ to the problem

$$\text{MIN } f_1(\underline{x})$$

$$\text{s.t. } \underline{x} \in T$$

and setting $f_{2MAX} = f_2(\underline{x}^*)$. Any value of $\underline{x}$ which gave $f_2(\underline{x}) > f_2(\underline{x}^*)$ was inferior since $f_1(\underline{x}) \geq f_1(\underline{x}^*)$ by definition of $\underline{x}^*$. For three or more objectives, however, a value of $\underline{x}$ which gives $f_2(\underline{x}) > f_2(\underline{x}^*)$ is not necessarily inferior since even though $f_1(\underline{x}) \geq f_1(\underline{x}^*)$ is still true, $f_j(\underline{x}) < f_j(\underline{x}^*)$ may hold for some j greater than 2. Thus $f_2(\underline{x}^*)$ is no longer the maximum value for f_2. However, there are ways of finding maximum values for certain problems. In many problems, the constraints will determine a maximum feasible value for each objective f_j which can be viewed as an upper bound for ε_j. In other words, the feasible set S in the functional space is bounded in all directions.

In other problems, one can determine a maximum value of any objective f_j at fixed values of the other objectives f_k, $k = 2,3,...n$, $k \neq j$. Fixing the values of these n-2 objectives reduces the problem to a two-dimensional one, so that the maximum value of f_j can be found by solving:

$$MIN \quad f_1(\underline{x})$$

$$s.t. \quad f_k(\underline{x}) \leq \varepsilon_k \qquad k = 2,3, ...,n, \quad k \neq j$$

$$\underline{x} \varepsilon T$$

If the solution to this problem is $\underline{x}^*$, and all of the ε_k constraints are binding, then the maximum value of f_j at these fixed values of other objectives is $f_j(\underline{x}^*)$. Note that for different values of the other objectives, there is no guarantee that this will still be the maximum value of f_j; this problem would have to be resolved with the new ε_k in order to find the new maximum value for f_j. In addition, this approach may prove quite difficult to implement.

The minimum value for each objective f_j can still be determined. It will be the solution to the problem

$$MIN \quad f_j(\underline{x})$$

$$s.t. \quad \underline{x} \varepsilon \dot{T}$$

6.3.2   Reversion to the Decision Space

The reversion to the decision space can be performed analogously to the two objective case. The preferred decision vector $\underline{x}^*$ and preferred value f_1^* of objective f_1 can be found as described in section 3.5.3 by solving the following problem:

Problem 6-2:

$$MIN \quad f_1(\underline{x})$$

$$\text{s.t.} \quad \underline{f}(\underline{x}) \leq \underline{f}^*$$

$$\underline{x} \in T$$

The alternative approach is as described in section 3.5.2, namely:

Problem 6-3: Solve the n-1 simultaneous equations:

$$\underline{\Lambda}_1(\underline{f}(\underline{x})) = \underline{\Lambda}_1^* \text{ such that } \underline{x} \text{ meets the Kuhn-Tucker conditions for problem 6-1.}$$

Theorem 6-4: Any solution $\underline{x}^*$ to the problem 6-2 also solves problem 6-3.

Proof: Assume $\underline{\Lambda}_1(\underline{f})$ is known. Then $\underline{\Lambda}_1^* = \underline{\Lambda}_1(\underline{f}^*)$. Since all the constraints $\underline{f}(\underline{x}) \leq \underline{f}^*$ must be binding (their multipliers are the components of $\underline{\Lambda}_1^*$ which are all greater than zero), then $\underline{f}(\underline{x}^*) = \underline{f}^*$. Thus $\underline{\Lambda}_1(\underline{f}(\underline{x}^*)) = \underline{\Lambda}_1^*$ so $\underline{x}^*$ would be found by problem 6-3.

Again $\underline{x}^*$ automatically satisfies the Kuhn-Tucker conditions since it solves the minimization problem 6-2, which is identical to problem 6-1 for the specific value $\underline{\varepsilon} = \underline{f}^*$. Q.E.D.

Thus it is not necessary to know $\underline{\Lambda}_1(\underline{f})$ in functional form. Note that the approach of solving problem 6-2 is generally simpler.

6.3.3   Multiple Regressions

Since λ_{1j} and f_1^* are now functions of $f_2, f_3, \ldots, f_n$, it is generally even more imperative than in the two objective case to avoid the multiple regressions which would be necessary to find their functional forms. Fortunately by solving problem 6-1 for any given value of $\underline{\varepsilon}$ which is binding, the values $\underline{\Lambda}_1(\underline{f})$ and $f_1^*(\underline{f})$ at $\underline{f} = \underline{\varepsilon}$ are found and thus the value of the worth functions $W_{1j}(\underline{f})$, $j = 2,3, \ldots n$, at $\underline{f} = \underline{\varepsilon}$ can be developed. Thus by solving problem 6-1 for q different values of $\underline{\varepsilon}$ which are binding, q different values of each W_{1j}, $j = 2,3,\ldots$, n, can be found without knowing the functional forms. The option of finding the functional forms in order to determine more values at which to ask the DM additional questions is still available, but the number of points necessary to get an accurate multiple regression will generally be inordinately large and will thus not be included in these algorithms. Alternatively, interpolation or curve fitting techniques can be used in place of regressions.

6.3.4   Finding the Indifference Band

The search techniques for finding the indifference band described in chapter four can be extended to the n-objective case, although now there are n-1 worth functions which must simultaneously equal zero.

For the exhaustive search technique, one would question the DM to find values for the n-1 worth functions at equally spaced points in the function space. If none of these had all n-1 worth functions equal to zero, then other values near the ones closest to zero would be tried until the indifference band is found.   Some effort can be saved by not trying all of the equally spaced points on the first pass. Rather, as soon as a point is found where all of the worth functions are near zero, the search procedure should be restarted with smaller increments from that point. Again, it may be necessary to apply interpolation, curve-fitting or regression to known non-inferior values in order to have the information to question the DM.

A gradient approach can also be used to determine which function space value to try next.   The gradient approach requires the equivalent of information about derivatives.   Therefore, n-1 values of each of the n-1 worth functions are required to determine this information for an n-objective problem. For example, in a three objective problem, to determine what to try next after $\underline{f}^0 = (f_2^0, f_3^0,)^T$, one must find $W_{12} (f_2^0 + \Delta_2, f_3^0)$, $W_{13}(f_2^0 + \Delta_2, f_3^0)$, $W_{12}(f_2^0, f_3^0 + \Delta_3)$ and $W_{13}(f_2^0, f_3^0 + \Delta_3)$. Then

$$\underline{f}^1 = \underline{f}^0 - \underline{J}^{-1}(W_{12}(\underline{f}^0), W_{13}(\underline{f}^0))^T(W_{12}(\underline{f}^0), W_{13}(\underline{f}^0))^T$$

where $\underline{J} = \begin{pmatrix} \frac{1}{\Delta_2}[W_{12}(f_2^0+\Delta_2, f_3^0) - W_{12}(f_2^0, f_3^0)] & \frac{1}{\Delta_3}[W_{12}(f_2^0, f_3^0 + \Delta_3) - W_{12}(f_2^0, f_3^0)] \\ \frac{1}{\Delta_2}[W_{13}(f_2^0+\Delta_2, f_3^0) - W_{13}(f_2^0, f_3^0)] & \frac{1}{\Delta_3}[W_{13}(f_2^0, f_3^0+\Delta_3) - W_{13}(f_2^0, f_3^0)] \end{pmatrix}$

This approach may also require interpolations, curve-fitting, or re-gression when the DM must be questioned at non-inferior values for which the trade-off ratios are not known. Both the exhaustive search and gradi-ent approach are also applicable when the λ-space surrogate worth func-tions are used by replacing f_j with λ_{1j}.

6.4   THE STATIC n-OBJECTIVE ε-CONSTRAINT (SNE) ALGORITHM

Algorithms describing the use of the SWT method for n-objective prob-lems will now be presented. The first of these uses the ε-constraint ap-proach both for finding the non-inferior points and for reverting to the de-cision space.   The water resources problems in chapter 8 will be solved by this method and thus no example is presented here. A flowchart is provided in figure 6-1.

Step 1: Find the minimum value of f_j by solving:

MIN $f_j(\underline{x})$

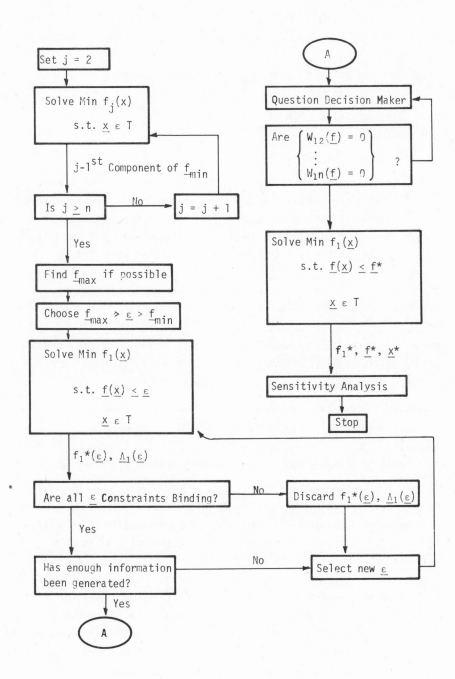

Figure 6-1. Flowchart for Static n-Objective ε-Constraint Algorithm.

s.t. $\underline{x} \in T$

The solution to this problem is the $j\text{-}1^{st}$ component of $\underline{f}_{MIN}$; this step is repeated for all $j = 2,3, \ldots,n$.

If possible the maximum values, $\underline{f}_{MAX}$ should be found here.

Step 2:   Set the initial values for $\underline{\varepsilon} > \underline{f}_{MIN}$

Step 3:   Solve   MIN $f_1(\underline{x})$

$$s.t.\ \underline{f}(\underline{x}) \le \underline{\varepsilon}$$
$$\underline{x} \in T$$

Let $\underline{x}^*$ be the decision vector which solves this problem. The solution is $f_1^*(\underline{\varepsilon}) = f_1(\underline{x}^*)$; each solution should also contain $\underline{\Lambda}_1(\underline{\varepsilon})$, the vector of Lagrange multipliers for the constraints.   If all of the $\underline{\varepsilon}$ constraints are binding then $\underline{\varepsilon} = \underline{f}$ so that the outputs of this step are $f_1^*(\underline{f})$ and $\underline{\Lambda}_1(\underline{f})$ at $\underline{f} = \underline{\varepsilon}$.   If any of the $\underline{\varepsilon}$ constraints are not binding then   ignore   these values.

Step 4:   If enough information   has been generated then proceed to step 5; otherwise select new values of $\underline{\varepsilon} > \underline{f}_{MIN}$ and return to step 3.   One method of selecting new values is to start with very large values for $\underline{\varepsilon}$ and decrease each ε_j by some number $\Delta_j > 0$ each iteration if the constraint $f_j(\underline{x}) \le \varepsilon_j$ is binding;   if this constraint is not binding then set $\varepsilon_j = f_j(\underline{x}^*)$.

If interaction with the   DM is possible on a real time basis, then the search techniques described in   section 6.3.4   can be used in choosing new values of $\underline{\varepsilon}$.

Step 5:   Develop the surrogate worth functions $W_{12}(\underline{f}),\ldots,W_{1n}(\underline{f})$ as follows: For each set of values $\underline{f}$, $\underline{\Lambda}_1(\underline{f})$, and $f_1^*(\underline{f})$ at which the worth   is desired, ask the   DM for his assessment of how much $\lambda_{1j}(\underline{f})$ additional units of objective f_1 are worth   in relation to one additional unit of objective f_j given $f_1^*(\underline{f})$ units of f_1 and the $j\text{-}1^{st}$ component of $\underline{f}$ units of f_j.   His assessment on a scale of   -10 to +10 with zero signifying equal preference is the value $W_{1j}(\underline{f})$.   This is repeated for all $j = 2,3, \ldots,n$.

Step 6:   Repeat step 5 until a value $\underline{f}^*$ is found such that all of the worth functions $W_{1j}(\underline{f}^*)$, $j = 2,3, \ldots,n$ equal zero.   Other values near $\underline{f}^*$ can be tried to determine the extent of the indifference band.

Step 7:   The preferred decision vector $\underline{x}^*$ is found by solving:

$$MIN\ f_1(\underline{x})$$
$$s.t.\ \underline{f}(\underline{x}) \le \underline{f}^*$$

$$\underline{x} \; \epsilon \; T$$

If there is more than one solution $\underline{f}^*$ to step 6, then this step must be re-peated for each one in order to find all of the preferred solutions.

Step 8: A sensitivity analysis should be performed to determine the possible effects of implementing the preferred solution.

Step 9: Stop!

6.5 THE STATIC n-OBJECTIVE MULTIPLIER (SNM) ALGORITHM

In order to determine the zeroes of the worth functions and thus the preferred solutions, it is generally necessary to have a large number of values for the worth functions; thus a large number of solutions to the minimizations in the first segment are required. Since the multiplier ap-proach generally is much more efficient than the ϵ-constraint approach in generating non-inferior points, it is useful in n-objective non-linear problems despite the possible inaccuracies caused by non-convexities (see section 4.3.1). For linear problems, the ϵ-constraint approach will not be too difficult due to the availability of the simplex method. The mixed algorithm will not be presented here; it can be generated by using the first three steps of this algorithm followed by steps 5 through 9 of the ϵ-constraint algorithm.

When there are more than two objectives, there is no simple means of finding λ_{1jMAX}. For this case it is assumed that $\lambda_{1jMAX} = \infty$ for j = 2,3, ... n. A flowchart of this algorithm is given in figure 6-2.

Step 1: Choose initial values for $\underline{\Lambda}_1 > 0$.

Step 2: Solve MIN $f_1(\underline{x}) + \underline{\Lambda}_1^T \cdot \underline{f}(\underline{x})$

s.t. $\underline{x} \; \epsilon \; T$

The solution vector $\underline{x}^*$ is substituted into $f_1(\underline{x})$ and $\underline{f}(\underline{x})$ to find $f_1^*(\underline{\Lambda}_1)$ and $\underline{f}^*(\underline{\Lambda}_1)$.

Step 3: If enough information has been generated, go on to step 4; if not, choose a new value for $\underline{\Lambda}_1 > 0$ and go back to step 2.

Step 4: Develop the surrogate worth functions $W_{12}(\underline{\Lambda}_1), \ldots, W_{1n}(\underline{\Lambda}_1)$ as follows: For each set of values $\underline{\Lambda}_1$, $f_1^*(\underline{\Lambda}_1), \underline{f}^*(\underline{\Lambda}_1)$ at which the worth is desired, ask the DM for his assessment of how much λ_{1j}, which is the j-1st component of $\underline{\Lambda}_1$, additional units of objective f_1 are worth in relation to one additional unit of objective f_j, given $f_1^*(\underline{\Lambda}_1)$ units of f_1 and the j-1st component of $\underline{f}^*(\underline{\Lambda}_1)$ units of f_j. His assessment on a scale of -10 to +10 is the value of $W_{1j}(\underline{\Lambda}_1)$. This step is repeated for all j = 2,3, ...,n.

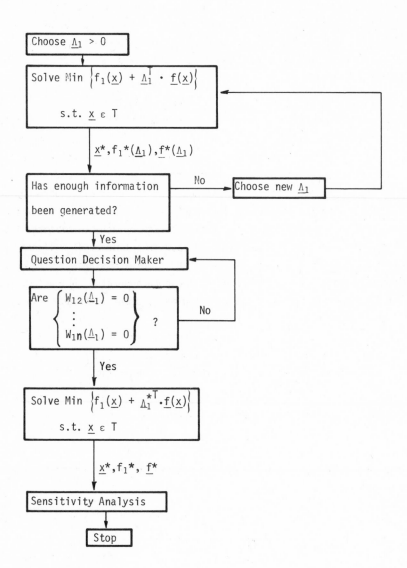

Figure 6-2. Flowchart for Static n-Objective Multiplier Algorithm.

Step 5: Repeat step 4 until $\underline{\Lambda}_1^*$ is found such that $W_{ij}(\underline{\Lambda}_1^*) = 0$ for $j = 2,3, \ldots n$. Additional values near $\underline{\Lambda}_1^*$ may be tried to determine the extent of the indifference band.

Step 6: Find the preferred decision vector $\underline{x}^*$ by solving

$$\text{MIN} \quad f_1(\underline{x}) + \underline{\Lambda}_1^{T*} \cdot \underline{f}(\underline{x})$$

$$\text{s.t.} \quad \underline{x} \, \epsilon \, T$$

Since this is the same problem as step 2 with $\underline{\Lambda}_1$ replaced by $\underline{\Lambda}_1^*$, little additional computation will be necessary.

Step 7: A sensitivity analysis could be performed to determine the possible effects of implementing the preferred solution.

Step 8: Stop!

6.6 SUMMARY

This chapter has extended the algorithms of chapter four for static n objective problems. The W_{1j} are found to be functions of $f_2, f_3, \ldots, f_n$ (or $\lambda_{12}, \lambda_{13}, \ldots, \lambda_{1n}$) and the preferred solution is defined to be that for which all of the W_{1j}, $j = 2,3, \ldots, n$ simultaneously equal zero. In addition, the use of the other worth functions W_{ij}, $i = 2,3, \ldots, n$, $j = 1,2, \ldots, n$, $j \neq i$, is shown to be redundant under certain assumptions. The computational efficiencies of avoiding regressions, finding maximum values for ϵ_j, using search techniques, and reversion to the decision space developed for two objective problems were modified for the n-objective case. Two algorithms were presented; the ϵ-constraint approach is useful for linear (and linearized) problems, but in other problems may require too much computational effort in order to generate enough values to have valid approximations to the worth functions. The multiplier approach is subject to inaccuracies due to non-convexities in the trade-off functions, but is generally much simpler to solve as it has n-1 fewer constraints. Mixed algorithms can also be used as a compromise between efficiency and accuracy.

The next chapter will discuss the modifications necessary for dynamic n-objective problems.

FOOTNOTES

1.   This new definition of $\underline{f}$ will be used throughout the rest of the book.

2.   See Feinberg [1972].

REFERENCES

1. Feinberg, A., "An Experimental Investigation of an Interactive Approach for Multi-criterion Optimization with An Application to Academic Resource Allocation," Western Management Science Institute, Working paper no. 186, 1972.

Chapter 7

THE SWT METHOD FOR DYNAMIC n-OBJECTIVE PROBLEMS

This chapter is both a modification of the previous chapter and an extension of chapter 5. Analysis of dynamic problems with more than two objectives and algorithms for their solutions are presented.

7.1 INTRODUCTORY ANALYSIS

For notational convenience define $\underline{\phi} \, \epsilon R^{n-1} = (\phi_2, \phi_3, \ldots, \phi_n)^T$ and $\underline{a} \, \epsilon \, R^{n-1} = (a_2, a_3, \ldots, a_n)^T$. A vector of new state variables $\underline{y} \, \epsilon \, R^{n-1}$ is defined such that $\dot{\underline{y}}(t) = \underline{a}(\underline{x}(t), \underline{u}(t), t)$ and $\underline{y}(0) = \underline{0}$; the remainder of the following notation is as defined in chapter 5. The problem in ϵ-constraint form then becomes:

Problem 7-1:   MIN $\phi_1(\underline{x}(t_f)) + \displaystyle\int_0^{t_f} a_1(\underline{x}(t), \underline{u}(t), t) \; dt$

$$\text{s.t.} \; \dot{\underline{x}}(t) = \; \underline{\psi}(\underline{x}(t), \underline{u}(t), t) \; ; \; \underline{x}(0) \; \text{given}$$

$$\dot{\underline{y}}(t) = \underline{a}(\underline{x}(t), \underline{u}(t), t) \; ; \; \underline{y}(0) \; = \; \underline{0}$$

$$\underline{g}(\underline{x}(t_f), t_f) \leqslant \; \underline{0}$$

$$\underline{y}(t_f) + \; \underline{\phi}(\underline{x}(t_f)) \leqslant \underline{\epsilon}$$

Note that again the objective functions are scalar valued since they are integrals over time. Also note that problems where t_f is a control variable (e.g. minimum time problems), as well as those with path constraints $\underline{g}(\underline{x}(t)\underline{u}(t), t) \leqslant \underline{0}$ can also be handled with the algorithms described in this chapter by modifying the necessary conditions for a minimum.[1]

Since the functional space for the dynamic problem is the same as the functional space for static ones (R^n), all analysis in the preceeding chapter which was conducted in the functional space will be applicable to this chapter. In review, the worth functions W_{ij}, i = 1,2, ...,n, j = 1,2, ..., n, i $\neq$ j are functions of $f_2, f_3, \ldots, f_n$ or $\lambda_{12}, \lambda_{13}, \ldots, \lambda_{1n}$, and the preferred solution $(f_1^*; \underline{f}^*)$ is where all $W_{1j}(\underline{f}) = 0$ for j = 2,3, ...,n simultaneously. The two assumptions of chapter 6 will still be maintained. They are that the trade-off rate is a good approximation to the change which would occur in the non-inferior value of f_i when f_j is changed by one unit, and that the DM is able to accurately assess his preferences. Then, theorems 6-1 through 6-3 will still hold. The following definition will also be made.

Definition 7-1: A preferred control vector and a preferred state vector are
defined as any feasible control vector $\underline{u}^*(t)$ and any feasible state vector
$\underline{x}^*(t)$ such that

$$\phi_1(\underline{x}^*(t_f)) \;+\; \int_0^{t_f} a_1(\underline{x}^*(t),\underline{u}^*(t),t)\ dt \;=\; f_1^* \quad \text{and}$$

$$\underline{\phi}(\underline{x}^*(t_f)) \;+\; \int_0^{t_f} \underline{a}(\underline{x}^*(t),\underline{u}^*(t),t)\ dt \;=\; \underline{f}^*$$

The computational efficiencies in chapter 6 are also applicable to
dynamic problems.   In particular, minimum (and sometimes maximum) values
for each f_j can be found, and search techniques can be used to find where
all of the worth functions equal zero.   Also, multiple regressions   or
interpolations to find $\underline{\Lambda}_1$ and f_1^* at other values of $\underline{f}$ may be used to avoid
resolving problem 7-1.   In addition, the preferred control and state vec-
tors can be found analogously to section 6.3.2.

7.2   THE DYNAMIC n-OBJECTIVE ε-CONSTRAINT (DNE) ALGORITHM

An algorithm will now be presented which uses the ε-constraint approach for both finding the non-inferior points and for reverting to the
decision space. A flowchart is provided in figure 7-1.

Step 1: Find the minimum value of f_j by solving:

$$\text{MIN} \quad \phi_j(\underline{x}(t_f)) \;+\; \int_0^{t_f} a_j(\underline{x}(t),\underline{u}(t),t)\ dt$$

$$\text{s.t.} \quad \underline{\dot{x}}(t) \;=\; \underline{\psi}(\underline{x}(t),\underline{u}(t),t) \;\;;\;\; \underline{x}(0) \;\; \text{given}$$

$$\underline{g}(\underline{x}(t_f),t_f) \;\le\; \underline{0}$$

The solution to this problem is the j-1st component of $\underline{f}_{MIN}$; this step is
repeated for all $j = 2,3, \ldots, n$. If possible the maximum value of each f_j,
$j = 2,3, \ldots, n$ should be found as described in section 6.3.1.

Step 2: Choose initial values for $\underline{\varepsilon} > \underline{f}_{MIN}$.

Step 3: Solve problem 7-1. The solution is $f_1^*(\underline{\varepsilon})$; each solution
should also contain $\underline{\Lambda}_1(\underline{\varepsilon})$, the vector of Lagrange multipliers for the $\underline{\varepsilon}$
constraints. If all of these contraints are binding, then $\underline{\varepsilon} = \underline{f}$ so that
the outputs of this step are $f_1^*(\underline{f})$ and $\underline{\Lambda}_1(\underline{f})$ at the value $\underline{f} = \underline{\varepsilon}$. If any of
the $\underline{\varepsilon}$ constraints are not binding then ignore these values.

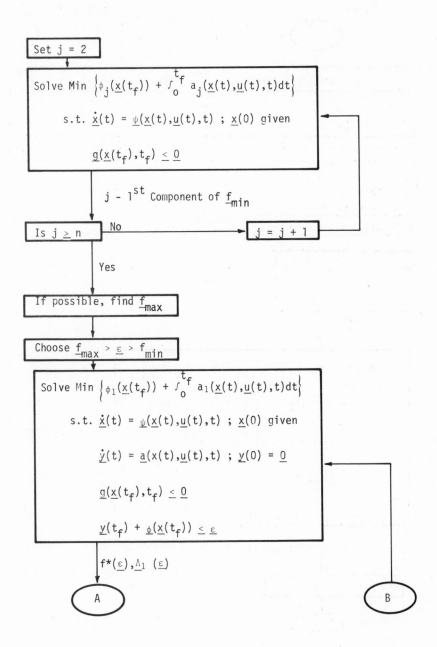

Figure 7-1. Flowchart for Dynamic n-Objective ε-Constraint Algorithm
Continued next page.

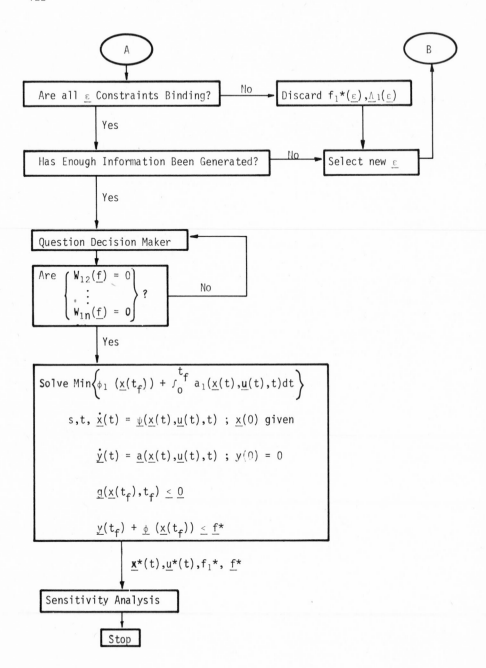

Figure 7-1 Continued.

Step 4: If enough information has been generated, then proceed to step 5; otherwise select new values for $\varepsilon > f_{MIN}$ and return to step 3.

Step 5: Develop the surrogate worth functions $W_{12}(\underline{f}),\ldots,W_{1n}(\underline{f})$ as follows. For each set of values $\underline{f}$, $\Lambda_1(\underline{f})$, $f_1^*(\underline{f})$ at which the worth is desired, ask the DM for his assessment of how much $\lambda_{1j}(\underline{f})$ additional units of objective f_1 are worth in relation to one additional unit of objective f_j, given $f_1^*(\underline{f})$ units of f_1 and the $j-1^{st}$ component of $\underline{f}$ units of f_j. His assessment on a scale of -10 to +10 is the value $W_{1j}(\underline{f})$. This is repeated for all $j = 2,3, \ldots, n$.

Step 6: Repeat step 5 until $\underline{f}^*$ is found such that $W_{1j}(\underline{f}^*) = 0$ for $j = 2,3, \ldots, n$. Additional values near $\underline{f}^*$ may be tried to determine the extent of the indifference band.

Step 7: Find the preferred state vector $\underline{x}^*(t)$ and control vector $\underline{u}^*(t)$ by solving problem 7-1 with ε replaced by $\underline{f}^*$. If there is more than one solution $\underline{f}^*$ to step 6, then this step must be repeated for each one in order to find all of the preferred solutions.

Step 8: A sensitivity analysis could be performed here to determine the possible effects of implementing the preferred solution.

Step 9: Stop!

7.3 THE DYNAMIC n-OBJECTIVE MULTIPLIER (DNM) ALGORITHM

Just as in the static case, finding the zeroes of the worth functions induces a need for a large number of solutions to the minimizations in the first segment. Usually the multiplier approach requires much less computation per solution than the ε-constraint approach, and its inaccuracies should be small compared with the inaccuracies in regression or interpolation; thus it is often best for non-linear problems. The mixed algorithm for dynamic problems will not be presented here; it can be generated by using the first three steps of this algorithm followed by steps 5 through 9 of the previous algorithm. A flowchart of this algorithm is given in figure 7-2.

Step 1: Set initial values for $\Lambda_1 > \underline{0}$.

Step 2: Solve the following problem:

$$\text{MIN} \quad \phi_1(\underline{x}(t_f)) + \Lambda_1^T \cdot \underline{\phi}(\underline{x}(t_f)) + \int_0^{t_f} \{a_1(\underline{x}(t),\underline{u}(t),t) +$$

$$\Lambda_1^T \cdot \underline{a}(\underline{x}(t),\underline{u}(t),t)\} \; dt$$

$$\text{s.t.} \quad \underline{\dot{x}}(t) = \underline{\psi}(\underline{x}(t),\underline{u}(t),t) \; ; \; \underline{x}(0) \; \text{given}$$

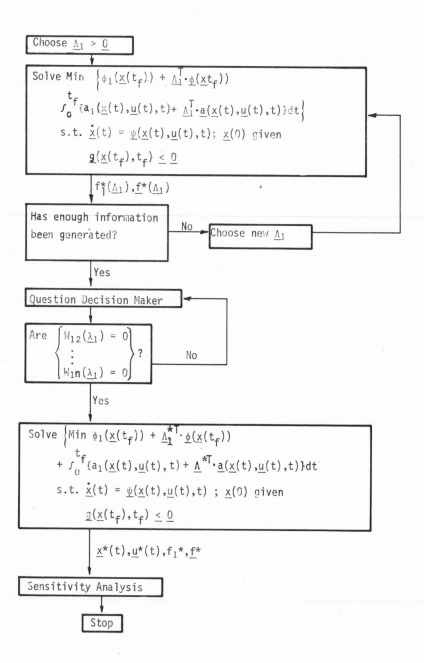

Figure 7-2. Flowchart for Dynamic n-Objective Multiplier Algorithm

$$\underline{g}(\underline{x}(t_f), t_f) \leq \underline{0}$$

The solution state vector $\underline{x}^*(t)$ and control vector $\underline{u}^*(t)$ are substituted

into $f_1 = \phi_1(\underline{x}(t_f)) + \int_0^{t_f} a_1(\underline{x}(t), \underline{u}(t), t) \, dt$ to get $f_1^*(\underline{\Lambda}_1)$ and into

$\underline{f} = \underline{\phi}(\underline{x}(t_f)) + \int_0^{t_f} \underline{a}(\underline{x}(t), \underline{u}(t), t) \, dt$ to get $\underline{f}^*(\underline{\Lambda}_1)$.

Step 3: If enough information has been generated, proceed to step 4; otherwise choose a new value $\underline{\Lambda}_1 > \underline{0}$ and return to step 2.

Step 4: Develop the surrogate worth functions $W_{12}(\underline{\Lambda}_1), \ldots, W_{1n}(\underline{\Lambda}_1)$ as follows. For each set of values $\underline{\Lambda}_1$, $f_1^*(\underline{\Lambda}_1), \underline{f}^*(\underline{\Lambda}_1)$ at which the worth is desired, ask the DM for his assessment of how much λ_{1j} additional units of objective f_1 are worth in relation to one additional unit of f_j given $f_1^*(\underline{\Lambda}_1)$ units of f_1 and the j-1st component of $\underline{f}^*(\underline{\Lambda}_1)$ units of f_j. His assessment on a scale of -10 to +10 is the value $W_{1j}(\underline{\Lambda}_1)$. This step is repeated for all $j = 2, 3, \ldots, n$.

Step 5: Repeat step 4 until $\underline{\Lambda}_1^*$ is found such that $W_{1j}(\underline{\Lambda}_1^*) = 0$ for $j = 2, 3, \ldots, n$. Additional values near $\underline{\Lambda}_1^*$ may be tried to determine the extent of the indifference band.

Step 6: Find the preferred state vector $\underline{x}^*(t)$ and preferred control vector $\underline{u}^*(t)$ by solving the problem in step 2 with $\underline{\Lambda}_1$ replaced by $\underline{\Lambda}_1^*$. If there was more than one preferred solution $\underline{\Lambda}_1^*$ in step 5, then this step is repeated for each one in order to find all of the preferred solutions.

Step 7: A sensitivity analysis could be performed to determine the possible effects of implementing the preferred solution.

Step 8: Stop!

7.4 SUMMARY

This chapter has extended the algorithms of chapter 5, and modified the results of chapter 6 for dynamic n-objective problems. The ε-constraint approach is useful for linear (and linearized) problems, but in other problems may require too much computation in order to generate enough values to have valid approximations to the worth functions. The multiplier approach is subject to inaccuracies due to non-convexities in the trade-off functions, but is generally much simpler to solve as it has n-1 less constraints. Mixed algorithms can also be used as a compromise between efficiency and accuracy.

The next chapters will provide examples of the use of the SWT method in problems in water resource planning.

FOOTNOTES

1. A derivation of the necessary conditions for a minimum in dynamic problems can be found in Bryson and Ho [1969] or other books on optimal control.

REFERENCES

1. Bryson, A.E., and Ho, Y.C., Applied Optimal Control. Ginn and Co., Waltham, Mass., 1969.

Chapter 8

APPLICATIONS OF THE SWT METHOD TO WATER RESOURCES PROBLEMS

8.1 THE REID-VEMURI EXAMPLE PROBLEM

Reid and Vemuri[1] introduced the following multiobjective function problem in water resources planning:

> "... A dam of finite height impounds water in the reservoir and that water is required to be released for various purposes such as flood control, irrigation, industrial and urban use, and power generation. The reservoir may also be used for fish and wildlife enhancement, recreation, salinity and pollution control, mandatory releases to satisfy riparian rights of downstream users and so forth. The problem is essentially one of determining the storage capacity of the reservoir so as to maximize the net benefits accrued..."

There are two decision variables:

x_1 = Total man hours devoted to building the dam.

x_2 = Mean radius of the lake impounded in some fashion.

There are three objective functions:

$f_1(x_1,x_2)$ = Capital cost of the project.

$f_2(x_2)$ = The water loss (volume/year) due to evaporation.

$\hat{f}_3(x_1,x_2)$ = The total volume capacity of the reservoir.

In order to change the volume objective to a minimization problem, the reciprocal function $f_3(x_1,x_2)$ was formed:

$$f_3(x_1,x_2) = 1/\hat{f}_3(x_1,x_2)$$

where

$$f_1(x_1,x_2) = e^{0.01x_1}(x_1)^{0.02}(x_2)^2$$

$$f_2(x_2) = 1/2\ (x_2)^2$$

$$f_3(x_1,x_2) = e^{-0.005x_1}(x_1)^{-0.01}(x_2)^{-2}$$

All decisions and objectives are constrained to be non-negative. Although this problem is far from representing a realistic water resource problem (there are only two decision variables), it was chosen because of the general interest that Reid and Vemuri had generated by their paper.

Reid and Vemuri were satisfied with determining non-inferior solutions via their proposed approach. With the SWT method, not only the same non-inferior solutions (in addition to others) can be generated, but also the trade-off rate functions and the ultimate preferred solution to the whole problem can be determined assuming the existence of a decision maker.

8.2   SOLUTION TO THE REID-VEMURI PROBLEM

This solution procedure will illustrate the approach of considering the surrogate worth functions as functions of the λ_{ij} described in section 3.5.2. The first step of the SWT method is to find the minimum values for each objective function. Clearly $\bar{f}_1 = 0$, $\bar{f}_2 = 0$ at $x_2 = 0$, and $\bar{f}_3 = 0$ at $x_1 = \infty$. The ε-constraint formulation is now adopted to generate λ_{12} and λ_{13}:

Problem 8-1:

$$\text{MIN} \qquad e^{.01x_1}(x_1)^{.02}(x_2)^2$$

$$\text{subject to} \quad 1/2\ x_2^2 \le \varepsilon_2$$

$$e^{-0.005x_1}(x_1)^{-0.01} x_2^{-2} \le \varepsilon_3$$

$$x_1 \ge 0, \quad x_2 \ge 0 .$$

Then the Lagrangian is formed:

$$L = e^{.01x_1}(x_1)^{.02} x_2^2 + \lambda_{12}(1/2\ x_2^2 - \varepsilon_2) + \lambda_{13}(e^{-.005x_1}(x_1)^{-.01} x_2^{-2} - \varepsilon_3)$$

The Kuhn-Tucker necessary conditions for a minimum are:

$$x_i \frac{\partial L}{\partial x_i} = 0; \quad \frac{\partial L}{\partial x_i} \ge 0 ; \quad x_i \ge 0,\ i = 1,2$$

$$\lambda_{ij} \frac{\partial L}{\partial \lambda_{ij}} = 0 ; \quad \frac{\partial L}{\partial \lambda_{ij}} \le 0 ; \quad \lambda_{ij} \ge 0 ; \quad j = 1,2$$

The above conditions were solved for various values of ε_2 and ε_3 (including some of the values from Table 1, Reid and Vemuri)[2] via the Newton-Raphson Method. The results are presented in Table 8-1.

The first two columns of Table 8-1 are the selected values of ε_2 and ε_3 (or equivalently f_2 and f_3). Columns three and four are the non-inferior values of the decision variables corresponding to the chosen values for ε_2 and ε_3. Column five is the corresponding non-inferior value of objective f_1. Columns six and seven are the trade-off ratios. Note that column seven is the ratio $\hat{\lambda}_{13} = -\partial f_1/\partial \hat{f}_3$. This is required in place of $\lambda_{13} = -\partial f_1/\partial f_3$ since the decision maker is familiar with the volume of the reservoir, $\hat{f}_3(x_1,x_2)$, rather than its reciprocal $f_3(x_1,x_2)$. This trade-off ratio can be found as follows:

$$\hat{\lambda}_{13} = - \frac{\partial f_1}{\partial \hat{f}_3} = - \frac{\partial f_1}{\partial f_3} \frac{df_3}{d\hat{f}_3}$$

$$= \lambda_{13} (- \frac{1}{(\hat{f}_3)^2}) = - \frac{\lambda_{13}}{(\hat{f}_3)^2}$$

An attempt to use multiple regression analysis for the construction of λ_{12} and $\hat{\lambda}_{13}$ as analytic functions of f_2 and f_3 using the wide band of non-inferior points (see Table 8-1) resulted in correlation coefficient of only .80. This is attributed to the exponential nature of the objective functions. Consequently, the alternative approach of avoiding regressions was adopted, where the decision maker provided the surrogate worth values W_{12} and W_{13}, for those values of λ_{12} and $\hat{\lambda}_{13}$, given in Table 8-1. The corresponding f_1, f_2, and f_3 can also be found in the table. If it is necessary to determine the worth at values of λ_{12} and $\hat{\lambda}_{13}$ not in Table 8-1, then interpolation or multiple regression near the desired values can be used.

The values of the surrogate worth functions generated with a "decision maker" are tabulated in columns 8 and 9 of Table 8-1. Note that more than one set of trade-offs resulted in an indifference band, $W_{ij} = 0$. The corresponding values of λ_{12}, $\hat{\lambda}_{13}$, f_1, f_2, and f_3 can be read directly from Table 8-1, rows 9, 25, 30, and 32. All solutions corresponding to these rows are preferred; they are non-inferior solutions which belong to the indifference band.

The decision variables corresponding to the above preferred solutions can be obtained in several ways. The simplest way in this example is to use Table 8-1. Thus, for example, row 9 provides the following optimal decisions and values of the objective functions:

x_1 = 172.95; $\quad$ x_2 = 38.73 ; $\quad$ f_1 = 9374.98 ;

f_2 = 750.00; $\quad$ f_3 = 3750.00

In other problems, the methods for reverting to the decision space described in section 6.3.2 may be required.

TABLE 8-1

NON-INFERIOR POINTS AND DECISION MAKER RESPONSES

	f_2	f_3	x_1	x_2	f_1	λ_{12}	$\hat{\lambda}_{13}$	w_{12}	w_{13}
1	250.00	500.00	0.70	22.36	499.95	2.00	- 2.00	+ 8	+ 6
2	250.00	1000.00	128.91	22.36	2000.00	8.00	- 4.00	+ 2	+ 2
3	250.00	1750.00	239.59	22.36	6124.45	24.50	- 7.30	- 2	- 2
4	250.00	2500.00	310.41	22.36	12499.99	50.00	-10.00	- 5	- 5
5	250.00	3750.00	391.04	22.36	28124.09	112.49	-15.00	-10	-10
6	250.00	5000.00	448.28	22.36	49984.46	199.88	-19.99	-10	-10
7	750.00	1750.00	24.43	38.73	2041.46	2.72	- 2.33	+ 7	+ 5
8	750.00	2500.00	93.09	38.73	4166.41	5.55	- 3.33	+ 4	+ 3
9	750.00	3750.00	172.95	38.73	9374.98	12.50	- 5.00	0	0
10	750.00	5000.00	229.91	38.73	16665.71	22.22	- 6.67	- 2	- 2
11	100.00	1750.00	421.71	14.14	15310.72	153.09	-17.50	-10	-10
12	500.00	1750.00	102.65	31.62	3062.14	6.12	- 3.50	+ 4	+ 3
13	100.00	3750.00	573.53	14.14	70310.77	703.09	-37.50	-10	-10
14	500.00	3750.00	253.27	31.62	14060.19	28.12	- 7.50	- 3	- 3
15	1000.00	3750.00	116.19	44.72	7029.45	7.03	- 3.75	+ 3	+ 2
16	106.00	473.00	150.47	14.56	1055.33	9.96	- 4.46	0	+ 1
17	33.40	150.00	151.74	8.17	336.83	10.08	- 4.49	0	+ 1
18	334.00	1500.00	151.74	25.85	3368.26	10.08	- 4.49	0	+ 1
19	1060.00	4730.00	150.47	46.04	10553.25	9.96	- 4.46	0	+ 1
20	31.60	316.00	310.41	7.95	1580.00	50.00	-10.00	- 5	- 5
21	3.34	150.00	609.47	2.58	3367.91	1008.25	-44.90	-10	-10

Table 8-1 (Cont'd)   Non-Inferior Points and Decision Maker Responses

	f_2	f_3	x_{13}	x_2	f_1	λ_{12}	$\hat{\lambda}_{13}$	W_{12}	W_{13}
22	59.50	841.00	379.22	10.91	5943.42	99.89	-14.13	-10	- 9
23	88.90	562.00	219.38	13.33	1776.34	19.98	- 6.32	- 2	- 1
24	33.40	1500.00	609.47	8.17	33679.12	1008.25	-44.90	-10	-10
25	100.00	500.00	172.95	14.14	1250.00	12.50	- 5.00	0	0
26	100.00	1000.00	31C.41	14.14	5000.00	50.00	-10.00	- 5	- 5
27	100.00	5000.00	630.86	14.14	124971.65	1249.43	-49.98	-10	-10
28	500.00	1000.00	0.70	31.62	999.89	2.00	- 2.00	+8	+ 6
29	500.00	5000.00	310.41	31.62	24999.99	50.00	-10.00	- 5	- 5
30	1000.00	5000.00	172.95	44.72	12499.97	12.50	- 5.00	0	0
31	100.00	2500.00	492.62	14.14	31209.63	311.69	-24.95	-10	-10
32	500.00	2500.00	172.95	31.62	6249.99	12.50	- 5.00	0	0
33	1000.00	2500.00	37.39	44.72	3125.00	3.12	- 2.50	+7	+ 5

8.3   DISCUSSION OF RESULTS

In general, one may need additional analysis in the case where there
is no row with both W_{12} and W_{13} equal to zero.   In this case, a multiple
regression or interpolations can be conducted to obtain W_{12} and W_{13} each as
a function of λ_{12} and $\hat{\lambda}_{13}$ near values where the worth functions are close
to zero.   Then one would have to solve simultaneously the equations for
$W_{12}(\lambda_{12},\hat{\lambda}_{13}) = 0$ and $W_{13}(\lambda_{12},\hat{\lambda}_{13}) = 0$ to obtain estimates   for λ_{12}^{*} and $\hat{\lambda}_{13}^{*}$.
Similarly, a multiple regression or interpolations can be conducted for f_2
and f_3 as functions of λ_{12} and $\hat{\lambda}_{13}$ in order to provide the necessary information to the DM. Finally, one would solve problem 8-1 for $\varepsilon_2^{*} = f_2(\lambda_{12}^{*},\hat{\lambda}_{13}^{*})$
and $\varepsilon_3^{*} = f_3^{*}(\lambda_{12}^{*},\hat{\lambda}_{13}^{*})$ as described in section 6.3.2.   These complications
were avoided in this example since four values were found for which both
surrogate worth functions were zero.   The next example will provide an illustration of the case where no preferred solution is found in the table.
Note that the worth functions W_{23},W_{21},W_{31} and W_{32} were not found since it
was assumed that the DM is accurately assessing his preferences.

8.4   STREAM RESOURCE ALLOCATION PROBLEM

This section will demonstrate the use of the surrogate worth tradeoff
method in another water resources problem. The problem - the allocation of
stream resources - has been studied extensively from the single objective
viewpoint[3].   The static n-objective ε-constraint algorithm described in
Chapter 6 will be used to solve this as a multiple objective problem.

For illustrative purposes, consider a physical system consisting of
a reservoir upstream of a series of n municipal and industrial users dis-
charging into the river.   The reservoir   is   used for water supply
and can also be used for low flow augmentation (releasing of water downstream to dilute wastes).   We assume the existence of a regional authority
which has the ability to control the amount of B.O.D. (biological oxygen
demand) discharged daily into the stream by each user (e.g. via effluent
limitations); the regional authority is also responsible for regulating
water release from the reservoir.

Thus the decision maker for our problem is the regional authority which
determines the effluent limitations (% of treatment) for each user, and the
amount of water to be released from the reservoir for low flow augmentation.
Define the decision variables as follows: x_i, i = 1,2, ..., n, is the percent treatment to be used at the ith treatment plant (expressed as a deci-
mal), and y is the amount of water (in units of F_0 where F_0 is the unaug-

mented initial flow in the stream) to be released from the reservoir for flow augmentation.

It is assumed that the objectives of the DM are threefold.

1: Minimize the total cost of waste treatment in the region (it is assumed that the cost of releasing water for flow augmentation is negligible since the dam is already present). The cost functions for each individual user as given by Hass will be used for the Miami River. These are quadratic in the percent B.O.D. removed:

$$Cost_i = 160.8 + 26.7q_i + (640.7 + 255.7q_i) (x_i - .45)^2$$

where q_i is the total waste water load generated by the ith user in million gallons per day. Thus the total cost to the region is the sum of the individual costs:

$$f_1 = \sum_{i=1}^{n} \alpha_i + \beta_i (x_i - .45)^2$$

where $\alpha_i = 160.8 + 26.7q_i$ and $\beta_i = 640.7 + 255.7q_i$.

2: Maximize the water in the reservoir available for water supply. This depends on evaporation, rainfall, capacity, etc. All of these factors are lumped into one parameter S , the amount of water available to be released (in units of F_0) for the period being studied. Then this second objective becomes:

$$\text{Maximize } f_2 = S - y$$

3: Minimize the pollution in the stream. It is assumed that the quality of the water immediately downstream of the final user (point A in figure 8-1) is important to the DM. For example, this area may be used for recreational purposes so that the quality must be maintained, or it may be part of another jurisdictional area so that to avoid legal problems the quality must be maintained. The measure of quality to be used is the concentration of dissolved oxygen, D.O., in milligrams per liter.

The basic Streeter-Phelps equation[4] is used to relate the amount of B.O.D. discharged into the river to the D.O. level at any point in the stream. Following the approach of Hass, the river is divided into n reaches (the point where each user discharges his wastes defines a new reach). Then the D.O. concentration at point A is:

$$D.O. = [\sum_{i=1}^{n} a_i x_i + b_1 y + (b_1 - c_1)]/[(c_2 - b_2) - b_2 y]$$

where $\quad a_i = w_i \sum\limits_{j=1}^{n} G_j R^n_{j+1} K^{j-1}_i$

$$b_1 = F_0(c_0 R^n_1 + \sum\limits_{j=1}^{n} S_j(1-e^{-r_j t_j}) R^n_{j+1})$$

$$b_2 = F_0 \eta$$

$$c_1 = w_0 \sum\limits_{j=1}^{n} G_j R^n_{j+1} K^{j-1}_1 + \sum\limits_{j=1}^{n} a_j - \sum\limits_{j=1}^{n} R^n_j c^T_j$$

$$+ \; S_j(1-e^{-r_j t_j}) R^n_{j+1} (F_j - F_0)$$

$$c_2 = (F_n - F_0)\eta$$

w_i = untreated B.O.D. load (lbs/day) of the ith user

$$G_j = K_j(e^{-k_j t_j} - e^{-r_j t_j})/(r_j - k_j)$$

$$R^i_j = \exp(-\sum\limits_{m=i}^{j} r_m t_m)$$

$$K^j_i = \exp(-\sum\limits_{m=i}^{j} k_m t_m)$$

F_i = unaugmented river flow at beginning of reach i (cf/day)

c_0 = initial concentration of D.O. in stream (mg/liter)

S_i = saturation level for O_2 in i^{th} reach (mg/liter)

r_j = reaeration coefficient in reach j

t_j = time of travel for reach j

k_j = deoxygenation coefficient for reach j

η = conversion factor from lbs/cf to mg/liter

c^T_j = D.O. content (lbs/day) of added flow in reach j

This is basically a restatement of equation A-14 of the appendix of Hass. Thus objective three becomes:

Maximize $\quad [\sum\limits_{i=1}^{n} a_i x_i + b_1 y + (b_1 - c_1)]/[(c_2 - b_2) - b_2 y]$

There are also constraints on this system. Legal constraints require that each user employ at least primary treatment (assumed to be 45%), and over 99% treatment is presently physically impossible. Thus

$$.45 \le x_i \le .99 \quad \text{for} \quad i = 1,2, \dots, n \, .$$

Similarly, the amount of flow augmentation must be between zero and the amount available (S = 3.47) so that $0 \le y \le 3.47$

Introducing the transformation $x_i' = x_i - .45$ for each i, the overall problem becomes:

Problem 8-2:

$$\text{MIN} \quad f_1 = \sum_{i=1}^{n} \alpha_i + \beta_i x_i'^2$$

$$\text{MAX} \quad f_2 = S - y$$

$$\text{MAX} \quad f_3 = [\sum_{i=1}^{n} a_i x_i' + .45 \sum_{i=1}^{n} a_i + b_1 y + (b_1 - c_1)]/$$

$$[(c_2 - b_2) - b_2 y]$$

$$\text{s.t.} \quad 0 \le x_i' \le .54$$

$$0 \le y \le S$$

8-5. SOLUTION OF STREAM RESOURCE ALLOCATION PROBLEM

The first task is to put problem 8-2 into standard form.   The second and third objectives are transferred into minimizations by using the fact that $\text{MIN} \ f_j = - \text{MIN}(- f_j)$. Thus define $f_2' = - f_2$ and $f_3' = - f_3$;   the algorithm can now be initiated.

Step 1: The minimum value of f_2' is found by solving

$$\text{MIN} \ y - S$$
$$\text{s.t.} \ 0 \le y \le S$$

Using the value S = 3.47 from Hass, f_{2MIN}' must equal -3.47 since the minimum obviously occurs at y = 0. For this problem the maximum value for f_2' can also be found. This obviously occurs at y = 3.47 and is $f_{2MAX}' = 0$.

The minimum value of f_3' is found by solving

$$\text{MIN} \ - [\ \sum_{i=1}^{n} a_i x_i' + .45 \sum_{i=1}^{n} a_i + b_1 y + (b_1 - c_1)]/[(c_2 - b_2) - b_2 y]$$

$$\text{s.t.} \quad 0 \le y \le S ; \quad 0 \le x_i' \le .54 \quad i = 2,3, \dots, n$$

It is obvious that the minimum value of f_3' is the point where the D.O. level (f_3) is maximum, which must occur where each $x_i' = .54$ and y = 3.47. This gives the value $f_{3MIN}' = - 7.97$.   Similarly, f_{3MAX}' occurs where y = 0 and each $x_i' = 0$; thus it is found that $f_{3MAX}' = -4.78$. Note that the values of the constants a_i, b_1, b_2, c_1, c_2 are developed from the data in Hass; these values are presented in Table 8-2. In summary, this step has provided the vectors $\underline{f}_{MIN} = \begin{pmatrix} -3.47 \\ -7.79 \end{pmatrix}$ and $\underline{f}_{MAX} = \begin{pmatrix} 0 \\ -4.78 \end{pmatrix}$

Step 2: Rewrite the problem in ε-constraint form:

$$\text{MIN} \sum_{i=1}^{n} \alpha_i + \beta_i x_i'^2$$

$$\text{s.t.} \quad y - S \leq \varepsilon_2$$

$$- \left[\sum_{i=1}^{n} a_i x_i' + .45 \sum_{i=1}^{n} a_i + b_1 y + (b_1 - c_1) \right] / [(c_2 - b_2) - b_2 y] \leq \varepsilon_3$$

$$0 \leq x_i' \leq .54 \quad \text{for} \quad i = 1, 2, \ldots, n$$

$$0 \leq y \leq 3.47$$

TABLE 8-2

Physical Constants for Stream Allocation Problem

i	a_i	q_i
1	3331.84	45.2
2	342.96	4.7
3	1539.69	4.2
4	886.02	3.6
5	73.83	0.5
6	172.06	1.2
7	189.40	0.8
8	433.07	0.6
9	199.94	0.5
10	1913.91	3.2
11	1741.59	8.4
12	722.94	2.7
13	238.59	0.6
14	3633.82	12.1
15	2266.56	8.4

$b_1 = 7650.31$ $c_1 = 1740.36$

$b_2 = -971.27$ $c_2 = 1930.14$

Bringing all of the constants over to the right hand side gives:

$$\text{MIN} \sum_{i=1}^{n} \alpha_i + \beta_i x_i'^2$$

$$\text{s.t.} \quad y \leq S + \varepsilon_2$$

$$- \left[\sum_{i=1}^{n} a_i x_i' + (b_1 - b_2 \varepsilon_3) y \right] \leq b_1 - c_1 + (c_2 - b_2) \varepsilon_3 + .45 \sum_{i=1}^{n} a_i$$

$$0 \leq x_i' \leq .54 \quad \text{for } i = 1, 2, \ldots, n$$

$$0 \leq y \leq 3.47$$

This problem can be easily solved by quadratic programming (QP). The values of the constants q_i used in finding α_i and β_i are also given in Table 8-2. As the right hand side of the ε_3 inequality is negative, in order to use quadratic programming the constraint must be changed to

$$\sum_{i=1}^{n} a_i x_i' + (b_1 - b_2 \varepsilon_3) y \geq (c_1 - b_1) - (c_2 - b_2) \varepsilon_3 - .45 \sum_{i=1}^{n} a_i$$

Let

$$Q_1 = b_1 - b_2 \varepsilon_3$$

$$Q_2 = S + \varepsilon_2$$

$$Q_3 = c_1 - b_1 - (c_2 - b_2) \varepsilon_3 - .45 \sum_{i=1}^{n} a_i$$

The problem then becomes:

Problem 8-3:

$$\text{MIN} \sum_{i=1}^{n} \alpha_i + \beta_i x_i'^2$$

$$\text{s.t.} \quad y \leq Q_2$$

$$\sum_{i=1}^{n} a_i x_i' + Q_1 y \geq Q_3$$

$$0 \leq x_i' \leq .54$$

$$0 \leq y \leq 3.47$$

The initial values for $\underline{\varepsilon}$ were chosen to be $\begin{pmatrix} -3.4 \\ -7.5 \end{pmatrix}$

Steps 3 & 4: A quadratic programming solution procedure, using an upper bounding technique[5] to eliminate the $x_i \leq .54$ and $y \leq 3.47$ constr-

aints, was written and implemented on the GE 4060 computer. This was sol-
ved for 33 different values of $\underline{\varepsilon}$ of which 30 were binding. The solution to
the quadratic programming problem includes the Lagrange multipliers λ_2 and
λ_3 for the constraints in problem 8-3. Notice that $\lambda_2 = -\partial f_1/\partial Q_2$ and $\lambda_3 =$
$+\partial f_1/\partial Q_3$ (the plus sign is present because λ_3 corresponds to a $\geq$ constraint). To find λ_{12} note that

$$\lambda_{12} = -\partial f_1/\partial f_2 = -\frac{\partial f_1}{\partial Q_2} \cdot \frac{dQ_2}{d\varepsilon_2} \cdot \frac{d\varepsilon_2}{df_2}$$

Since $Q_2 = S + \varepsilon_2$ and $\varepsilon_2 = f_2' = -f_2$ then $\lambda_{12} = -\frac{\partial f_1}{\partial Q_2} \cdot (-1) = -\lambda_2.$

It is appropriate that λ_{12} is negative (i.e., $-\partial f_1/\partial f_2$ is positive) since
if the amount available for supply increases, then the cost to achieve the
same quality will also increase (since the flow augmentation will be less,
more treatment will be required). To find λ_{13}, note that

$$\lambda_{13} = -\partial f_1/\partial f_3 = -\frac{\partial f_1}{\partial Q_1} \cdot \frac{dQ_1}{d\varepsilon_3} \cdot \frac{d\varepsilon_3}{df_3} - \frac{\partial f_1}{\partial Q_3} \cdot \frac{dQ_3}{d\varepsilon_3} \cdot \frac{d\varepsilon_3}{df_3}$$

Now, $\varepsilon_3 = f_3' = -f_3$, so $\frac{d\varepsilon_3}{df_3} = -1$; also $\frac{dQ_3}{d\varepsilon_3} = b_2 - c_2$ and $dQ_1/d\varepsilon_3 = -b_2$.
The only problem is finding $\partial f_1/\partial Q_1$. Note that the Lagrangian for problem
8-3 is:

$$L = f_1 + \lambda_2(y-Q_2) - \lambda_3 (\sum_{i=1}^{n} a_i x_i' + Q_1 y - Q_3)$$

Since $L = f_1$ at the optimum, $\partial f_1/\partial Q_1 = \partial L/\partial Q_1 = -\lambda_3 y.$

Thus $\lambda_{13} = b_2\lambda_3 y - \lambda_3(b_2 - c_2)$.

The first five columns of Table 8-3 show the results of these steps.
The units for f_2 were changed from units of F_0 to million cubic feet per
day and for λ_{12} from \$/units of F_0 to \$/million cubic feet per day.
Step 5: The DM is questioned and his responses are given in the last
two columns of Table 8-3.   The DM for this problem was a graduate student
in the water resources group.   In order to impart an understanding of the
system for which he was making decisions, the objectives were discussed and
the upper and lower bounds for each objective were presented along with
Figure 8-1. The description given was as follows:
"A possible operating point for the system is f_2 million cubic feet

of water available for supply, a D.O. level of f_3 mg/l at point A in the stream, and a cost of f_1^* \$/day. At this point, would you be willing to pay an additional $- \lambda_{12}$ \$/day in order to have 1 million more cubic feet of water available? Rate your willingness on a scale from -10 (totally unwilling) to +10 (totally willing) with zero signifying indifference. Similarly, rate your willingness to pay an additional $- \lambda_{13}$ dollars per day to raise the D.O. level 1 mg/liter."

The unit increments for f_2 (1 million cubic feet of water) and f_3 (1 mg/liter) were chosen to be large enough so that the DM would be able to perceive the difference. For example, the DM would probably consider the difference between 5.0 and 5.1 mg/liter to be negligible and thus questions about raising the D.O. level by .1 mg/l would have no significance. At the same time, the unit increments must be small compared to the absolute values of the objectives, since $\lambda_{1j} = - \Delta f_1/\Delta f_j$ is only the case in the limit, as the increments go to zero. Thus, if the unit increments are too large, λ_{12} is not a good approximation to the change $- \Delta f_1/\Delta f_2$. It is felt that the values chosen fit both criteria.

Step 6: Since none of the non-inferior values in Table 8-3 had both worth functions equal to zero, multiple regressions were performed to approximate $W_{12}(f_2, f_3)$ and $W_{13}(f_2, f_3)$. Due to the relative and subjective nature of the DM's responses, it is felt that a linear approximation is as good as any. The results of the regression were:

$$W_{12}(f_2, f_3) \approx 47.55 - .34 f_2 - 4.94 f_3$$
$$W_{13}(f_2, f_3) \approx 67.66 - .15 f_2 - 9.88 f_3$$

The correlation coefficients were $R^2 = .721$ and $R^2 = .868$ respectively. These two equations were solved simultaneously to give an estimate of the preferred values $f_2^* = 51.77$ and $f_3^* = 6.06$. The corresponding tradeoff ratios were found by interpolation to be $\lambda_{12}^* = -61.2$ and $\lambda_{13}^* = -1670$ and the corresponding non-inferior value of f_1 found by interpolation is $f_1^* = 5987$. When these values were given to the DM, he assigned $W_{12} = 0$ and $W_{13} = 0$. Note that instead of interpolations, the problem in step 3 could have been resolved to find the trade-off rates and non-inferior value of f_1 for f_2^* and f_3^*.

Step 7: The reversion to the decision space is performed by solving the quadratic programming problem in step 3 with ε_2 and ε_3 replaced by f_2^* and f_3^*. The results give $x_i^{'*}$, i = 1,2, ...,n which must be transformed back to x_i^*, and y^*. The preferred solution is:

TABLE 8-3

Non-inferior Points and DM Responses for Stream Allocation Problem

f_2	f_3	f_1^*	λ_{12}	λ_{13}	W_{12}	W_{13}
50.54	5.0	4994	- 0.67	- 12	+ 10	+ 10
40.43	5.5	4994	- 1.89	- 48	+ 10	+ 9
46.66	5.5	5073	-23.53	- 532	+ 9	+ 8
52.88	5.5	5287	-45.16	- 904	+ 7	+ 6
24.88	6.0	4995	- 2.43	- 98	+ 10	+ 10
31.10	6.0	5052	-15.92	- 588	+ 9	+ 5
38.88	6.0	5242	-32.78	-1075	+ 3	+ 4
46.66	6.0	5562	-49.64	-1422	0	+ 3
52.88	6.0	5912	-63.13	-1599	- 5	- 3
3.89	6.5	5031	- 9.30	- 652	+ 10	+ 3
15.55	6.5	5219	-22.91	-1412	+ 7	0
23.33	6.5	5432	-31.98	-1791	+ 5	- 1
31.10	6.5	5716	-41.05	-2067	+ 3	- 4
38.88	6.5	6070	-50.13	-2241	0	- 5
46.66	6.5	6496	-59.20	-2312	- 5	- 10
52.88	6.5	6887	-66.46	-2295	- 9	- 10
3.89	7.0	6074	-31.97	-3521	+ 10	- 1
15.55	7.0	6479	-37.49	-3630	+ 8	- 4
23.33	7.0	6785	-41.17	-3621	+ 4	- 5
31.10	7.0	7120	-44.85	-3547	+ 2	- 7
38.88	7.0	7483	-48.53	-3407	- 1	- 7
46.66	7.0	7874	-52.21	-3202	- 4	- 10
52.88	7.0	8208	-55.15	-2991	- 10	- 10
3.89	7.5	8552	-24.93	-6390	+ 10	- 10
15.55	7.5	8849	-25.95	-5848	+ 9	- 10
23.33	7.5	9053	-26.63	-5451	+ 7	- 10
31.10	7.5	9263	-27.30	-5026	+ 5	- 10
38.88	7.5	9478	-27.98	-4573	+ 1	- 10
46.66	7.5	9698	-28.66	-4092	- 3	- 10
52.88	7.5	9878	-29.21	-3687	- 7	- 10

$$\underline{x}^{*T} = (.47, .50, .70, .61, .48, .50, .51, .60, .52, .81, .62,$$
$$.60, .53, .72, .67)$$

$$y^* = 0.14$$

$$f_1^* = 5943$$

$$f_2^* = 51.77$$

$$f_3^* = 6.06$$

Step 8: At this point some sort of sensitivity analysis could be performed; however, such work is beyond the scope of this book. The only check made was to inform the DM of the results, and he felt that it did express his preferences fairly accurately.

8.6 DISCUSSION OF RESULTS

The surrogate worth trade-off method was discovered to be easily implementable for this problem. The mathematical model for this problem was already available from Hass[6], the only change necessary being that of enabling the value of ε_3 to change. This required only slight modification of that model. The fact that the problem was formulated in a manner permitting the use of quadratic programming greatly reduced the programming and debugging time. Quadratic programming uses the simplex method for which there are many packages available; the only specific modifications that must be made are for interfacing the data, which did not require too much effort to implement in this situation. In addition, the simplex technique is a very efficient procedure.

Thirty different non-inferior points were found using less than twenty minutes of computer time of the GE 4060 digital computer (which is equivalent to approximately two or three minutes on the Univac 1108 computer); each non-inferior point took approximately thirty seconds of computer time as did the three inferior (non-binding) solutions. The interaction with the decision maker required less than thirty minutes, but this time period appeared to be more than sufficient to allow him to adequately express his preferences. These preferences were then used in a multiple linear regression to find an estimate of the preferred solution; again there are numerous computer packages available to perform regressions so little effort is necessary. Interpolations were used to find the corresponding trade-off rates and the DM was questioned again. The reversion to the decision space required only one more iteration of the quadratic programming procedure. In summary, the overall implementation and solution via the SWT method was

easily accomplished. Despite the simplicity of the problem in this example
it is an approximation to real situations, and thus the surrogate worth
trade-off method appears to have great potential for use in real problems.

8.7   NORTHERN CALIFORNIA WATER SYSTEM

Another problem for which multiple objective analysis is appropriate
is that of reservoir operations for water and energy on a month to month
basis.   A large scale example of this problem considered as a single objective problem is the analysis of the Northern California water system[7]
which considered the monthly operations to produce firm water and power
economic returns from a multi reservoir system over a 50 year planning
horizon.   In reality there are at least three objectives which should be
considered in the scheduling of reservoir releases. These are the energy
made available, water made available, and cost of operation. The California water systems analysis attempted to commensurate these objectives, aug-
menting energy and water into monetary terms by using the price for which
they could be sold as a commensurating coefficient. However, in general,
this price would not be a constant but would depend on the amount of energy
and water available as well as on time. Rather than attempt a price evaluation of all possible combinations of water and energy at each of 600 time
periods, multiobjective analysis can be used as a much simpler approach.

The system modeled here is a simplification of the Trinity subsystem
in the Northern California water system.   There are two reservoirs in this
system. The first releases water through a power plant; the second can release water either for supply or downstream through another power plant and
to the ocean.   The storage capacity of the second reservoir is negligible
compared to the first so that its level can be considered constant on a
monthly basis.   In other words, all water coming into the second (lower)
reservoir in any period is either released to the river or used for supply
and additional energy (see Figure 8-1).

For this system, there are two decisions which must be made for each
planning period - namely the releases from the two reservoirs.   These releases are denoted x_i and r_i respectively. Note that in reality these values are the average releases for the ith period. If ten planning periods
are considered, as in the following approximation, there are then 20 decisions.

It is also assumed that the two reservoir operations are not independent. In other words, there is some agency responsible for the overall
operation of the system and this agency will determine the releases to

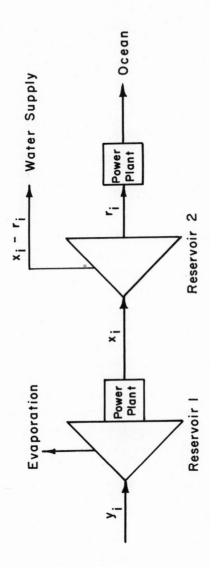

Figure 8-1:   A Two Reservoir System

optimally operate the system.

There are two ways of formulating the energy objective. The first recognizes that what is important to the region is the amount of energy which can be guaranteed during a planning period. Since industrial concerns or private utilities utilizing the energy will usually need to recover large amounts of capital over substantial periods of time to make use of the energy, there must be some reasonable assurance that the anticipated level of energy will be available as planned. This "guaranteed" energy is known as "firm" energy. Additional energy above the guaranteed amount ("dump" energy) is useful in that it can replace other sources of energy for the industries when it is available, but is somewhat less valuable than the guaranteed amount. If E_i is the amount of energy generated by the system in period i, then the firm energy can be written as Firm Energy = $\underset{i}{\text{Min}} \{E_i\}$, or $\underset{i}{\text{Min}}\{\alpha_i E_f\}$ where the fraction of the firm level E_f is α_i and is possibly different for each i. Note that in general, the utility is concerned with the amount of energy guaranteed per day rather than per month, but for this particular system energy contracts are in fact based on monthly demands during 'on peak' periods.

Another approach is to ignore the difference between firm and dump energy and assume that the total amount of energy produced is what is of value to the region regardless of when produced. This is a simplification which may or may not be justified for any particular region. If it is justified, the energy output can be written as Total Energy = $\sum_i E_i$.

The energy E_i will now be related to the decisions. Consider first the energy produced by a release of water from the first reservoir through the power plant. This energy would be proportional to the amount of water released, x_i, and also to the level of water in the reservoir. The level of the water in the reservoir, h, is a function of the volume of water stored, q, where generally it can be approximated by a function of the form $h = q^{1/3}$. Other more precise forms or even a tabulation can be used.

Note that the storage volume changes over time and is dependent on the inflows y_i and the amount of water released each period. Define q_i to be the storage volume at the beginning of period i. Note that the amount of storage at the end of period i is equal to the amount of storage at the beginning of period i + 1. The following equation is used to relate the storage volumes in successive periods.

$$q_{i+1} = q_i - x_i + y_i - e_i$$

where y_i is the inflow to and e_i is the evaporation from the reservoir in period i.

Although the inflow y_i is a stochastic parameter, the values of y_i will be assumed deterministic and known, and will be based on critical period hydrologic analysis. That is, the worst sequence values for y_i found in a 50-year sample of hydrologic records will be used. Since this "critical" period will control the maximum firm output for that particular hydrograph the results can be shown to be identical with the 10 period sequence. The evaporation from the reservoir in period i will in general be proportional to the surface area. This can usually be approximated by $e_i = \beta_i q_i^{2/3}$.

Since q changes between i and i+1, e_i will vary over time. The average value $\bar{e}_i$, over one period i can be approximated by

$$\bar{e}_i = \frac{1}{2} (e_i + e_{i+1}) = \frac{\beta_i}{2} (q_i^{2/3} + q_{i+1}^{2/3})$$

In general, the evaporation rate, β_i will vary with the season . Replacing this in the state equation for q_i gives

$$q_{i+1} + \frac{\beta_i}{2} q_{i+1}^{2/3} = q_i - \frac{\beta_i}{2} q_i^{2/3} - x_i + y_i$$

It will also be assumed that the initial storage in the reservoir, q_1, is known.

Returning to the calculation of the head, it is apparent that the average value of the head over the period must be used in determining the energy. Thus the average head, $\bar{h}$, is:

$$\bar{h} = \frac{1}{2} (h_i + h_{i+1}) = \frac{1}{2} (q_i^{1/3} + q_{i+1}^{1/3})$$

The energy output from the first reservoir can thus be written as:

$$E_i = \frac{1}{2} (q_i^{1/3} + q_{i+1}^{1/3}) x_i$$

The energy output from the second reservoir is much simpler to calculate since the head is constant due to a constant storage level. In addition it will be assumed that evaporation is negligible and that the inflow other than x_i is negligible. The energy from this reservoir is easily calculated as $E_i = \gamma r_i$. Therefore the total energy output of the system is the sum of these two energy outputs. In reality, there may be limits on the maximum capability of energy output for each of the power plants, but these are simple constraints which will be ignored in this analysis.

The objective of water supply also can be considered in two ways. The guaranteed "firm" water can be a separate objective, or the total water output over the ten period span can be the measure chosen. From Figure 8-1 it can be seen that the amount available for water supply, W_i, in any period is $W_i = x_i - r_i$.

The final objective is a cost function. It will be assumed that the cost of power generation at each plant is proportional to the amount of water passing through the power plant, and that the variable cost of water supply (largely pumping energy for distribution) is proportional to the amount of water supplied. Then the variable cost for the ith period can be written as

$$C_i = a_i x_i + b_i r_i + c_i (x_i - r_i) = (a_i + c_i) x_i + (b_i - c_i) r_i$$

In general, non-linear cost functions may be more appropriate but linearity in this instance is reasonable and will be assumed for simplicity without loss of generality insofar as multiple objective analysis is concerned.

A final constraint which must be considered for this system is that the amount of storage at the end of any period cannot exceed the capacity of the reservoir and must be above the minimum required for proper reservoir operation, $(Q_{min} \leq q_{i+1} \leq Q_{max})$.

The three objective optimization problem can be summarized as

Maximize firm energy: $\text{Max} \left[\underset{i}{\text{Min}} \left\{ \frac{1}{2} (q_i^{1/3} + q_{i+1}^{1/3}) x_i + \gamma r_i \right\} \right]$

Maximize firm water: $\text{Max} \left[\underset{i}{\text{Min}} \left\{ x_i - r_i \right\} \right]$

Minimize cost: $\text{Min} \left[\sum_i (a_i + c_i) x_i + (b_i - c_i) r_i \right]$

Subject to

$$q_{i+1} + \frac{\beta}{2} q_{i+1}^{2/3} = q_i - \frac{\beta}{2} q_i^{2/3} - x_i + y_i$$

$$x_i \geq r_i \geq 0$$

$$Q_{Min} \leq q_{i+1} \leq Q_{Max} \qquad i = 1, 2, \ldots, 10.$$

The alternative formulation replaces the first two objectives by

$$\text{Maximize total energy} = \text{MAX} \left[\sum_i \frac{1}{2} (q_i^{1/3} + q_{i+1}^{1/3}) x_i + \gamma r_i \right] .$$

$$\text{Maximize total water} = \text{MAX} \left[\sum_i x_i - r_i \right] .$$

Note that the cost objective does not change. To make this formulation more realistic, constraints requiring a minimum output of water and energy in each period could be included.

Both of these problems are highly non-linear and are difficult to solve. When put into ε-constraint form, the first formulation becomes:

$$\underset{x,r}{\text{Min}} \sum_{i=1}^{10} \{ (a_i + c_i) x_i + (b_i - c_i) r_i \}$$

s.t. 1) $q_{i+1} + \frac{\beta}{2} q_{i+1}^{2/3} = q_i - \frac{\beta}{2} q_i^{2/3} - x_i + y_i$

2) $Q_{min} \leq q_{i+1} \leq Q_{max}$

3) $\frac{1}{2} (q_i^{1/3} + q_{i+1}^{1/3}) x_i + \gamma r_i \geq \varepsilon_2$

4) $x_i - r_i \geq \varepsilon_3$

5) $r_i \geq 0$
   $i = 1,2,\ldots, 10.$

(Note that $x_i \geq 0$ is not necessary as a constraint since this condition must be met for constraints 4 and 5 to hold.)

With this formulation, there are 20 decision variables, one objective function, 10 equality constraints and 50 inequality constraints. The problem may be reformulated by using the equality constraints,

$$x_i = q_i - \frac{\beta}{2} (q_i^{2/3} + q_{i+1}^{2/3}) - q_{i+1} + y_i ,$$

and substituting for x.

The formulation becomes:

Problem 8-4:

$$\underset{i=1}{\text{Min}} \sum^{10} \{ (a_i + c_i)(q_i - \frac{\beta}{2} (q_i^{2/3} + q_{i+1}^{2/3}) - q_{i+1} + y_i)$$

$$+ (b_i - c_i) r_i \}$$

$$\text{s.t.} \quad 1) \quad \frac{1}{2}(q_i^{1/3} + q_{i+1}^{1/3})(q_i - \frac{\beta}{2}(q_i^{2/3} + q_{i+1}^{2/3}) - q_{i+1} + y_i)$$

$$+ \gamma r_i \geq \varepsilon_2$$

$$2) \quad q_i = \frac{\beta}{2}(q_i^{2/3} + q_{i+1}^{2/3}) - q_{i+1} + y_i - r_i \geq \varepsilon_3$$

$$(i = 1,2, \ldots, 10)$$

with bounds on the decision variables of

$$1) \quad Q_{min} \leq q_{i+1} \leq Q_{max}$$

$$2) \quad r_i \geq 0 \qquad\qquad (i = 1,2, \ldots, 10)$$

The decision variables are r_i and q_{i+1} for $i = 1,2, \ldots, 10$. (Remember that q_1 is a known constant.)

With this formulation there are 20 bounded decision variables, one objective function, and 20 inequality constraints.

8.8   SOLUTION OF CALIFORNIA WATER SYSTEM MULTIOBJECTIVE PROBLEM

The formulation involving cost, firm water, and firm energy as the three objectives was implemented and solved on the Univac 1108 Computer System. The values used for the constants are listed in Table 8-4. Finding the minimum and maximum values for ε_2 and ε_3 requires the solution of a non-linear programming problem with minimax objectives. In order to save programming time, an approximate lower limit of zero is used for both objectives. The maximum for f_2 is taken as 300 MWH since no feasible solution could be found for any higher values. Similarly, the maximum for f_3 is 7500 AF. This approximation may lead to some non-binding solutions (recognizable by $\lambda = 0$) which must then be discarded.

The minimization in problem 8-4 was solved for different values of ε_2 and ε_3 in the ranges defined in the preceeding paragraph. These values were chosen to be approximately equidistant in the feasible region and are listed in the first two columns of Table 8-5. Note that there are 10 different constraints for ε_2 in problem 8-4. In some cases, only one will be binding, and the corresponding non-zero Lagrange multiplier is the trade-off ratio λ_{12}. Should there be more than one non-zero multiplier, the one with the largest absolute value corresponds to the most binding constraint and thus represents the trade-off ratio. It is important to remember that λ_{12} describes the change in the optimal value of f_1 due to an incremental change in f_2 only for small increments. This is especially relevant here since a large increment may cause a different one of the ten constraints to become

TABLE 8-4

Data for Trinity River Subsystem Example

Period	y(KAF)	a($/AF)	b($/AF)	c($/AF)
1	18.0	3.00	2.00	0.60
2	22.5	3.00	2.00	0.70
3	25.0	5.00	2.50	0.60
4	30.0	3.00	2.00	0.50
5	27.5	2.50	1.50	0.50
6	15.0	2.00	1.00	0.40
7	10.0	1.00	0.50	0.30
8	10.0	1.50	1.00	0.40
9	15.0	2.50	2.00	0.50
10	17.0	4.00	2.50	0.60

α = 200 MWH/AF/Ft

β = .04 AF/Acre

γ = 1500 MWH/AF

q_1 = 180 KAF

Q_{min} = 100 KAF

Q_{max} = 400 KAF

TABLE 8-5

Results for Trinity River Subsystem Example

First run	f_2	f_3	f_1^*	λ_{12}	λ_{13}	W_{12}	W_{13}	
				Computational Model		Decision Maker		
	100	0	22	-.036	-3.6	+ 8	+10	
	100	2.5	82	-.028	-5.6	+ 9	- 2	
	150	0	32	-.036	-3.6	+ 4	+10	
	150	2.5	85	-.036	-4.6	+ 7	- 1	
	150	5.0	163	-.028	-5.6	+ 2	- 5	
	200	0	43	-.036	-3.6	+ 2	+10	
	200	2.5	95	-.037	-3.6	0	0	Preferred Solution
	200	5.0	165	-.028	-5.6	0	- 6	
	250	0	54	-.036	-3.6	- 4	+10	
	250	2.5	106	-.037	-3.6	- 1	0	
	250	5.0	166	-.028	-5.6	- 4	- 7	
	250	7.5	246	-.028	-5.6	- 3	-10	
	300	0	65	-.036	-3.5	- 6	+10	
	300	2.5	117	-.037	-3.6	- 7	- 1	
	300	5.0	170	-.036	-4.6	- 5	- 8	
	300	7.5	247	-.028	-5.6	- 4	-10	
Second run	220	2.5	99	-.037	-3.6	0	0	Preferred Solution
	180	2.5	91	-.037	-3.6	+ 2	0	
	200	3.0	106	-.037	-3.6	0	- 1	
	200	2.0	85	-.037	-3.6	0	+ 6	

Note: the units of f_1^* are \$1000

f_2 are MWH/period

f_3 are KAF/period

binding. Note also that $\lambda_{12} = -\partial f_1/\partial f_2$ will be negative, since an increase in cost (f_1) will yield an increase in firm energy (f_2). All of the above analysis also applies to λ_{13} and ε_3.

The solution procedure utilized for problem 8-4 is the generalized reduced gradient (GRG) algorithm for non-linear optimization[8]. A computer package is available for the Univac 1108 Computer System. Seventeen values of ε_2 and ε_3 were used as input, however one of them did not converge to a solution within the time limit of twenty-seconds, and four were inferior solutions. The solutions which did converge required, on the average, four seconds of computer time. The results of this computation provide the non-inferior value of f_1 and the trade-off ratios λ_{12} and λ_{13} corresponding to each pair $(\varepsilon_2, \varepsilon_3)$. Those results are listed in the middle three columns of Table 8-5.

The decision-maker (DM) was interrogated with questions analogous to those used for the stream resource allocation problem. For example, the question corresponding to row 2 in Table 8-5 "Given an operating policy costing $ 82,000 for the ten planning periods which guarantees 2500 acre feet of firm water per period and 100 megawatt hours of firm power per period, how willing would you be to spend an additional $2800 to increase the firm energy by 100 MWH/period? At the same point, what would your willingness be to spending an additional $5600 to increase the firm water by 1000 acre feet per period?" The responses of the DM are listed in the last two columns of table 8-5. One preferred solution is found in this table - namely, $f_1^* = \$95,000$, $f_2^* = 200$ MWH/period, $f_3^* = 2500$AF/period. Four other values of ε_2 and ε_3 near f_2^* and f_3^* were used in the GRG algorithm to try to determine the extent of the indifference band. These computations and results are listed in the last four rows of Table 8-5. It can be seen that the indifference band extends approximately from $200 \leq f_2^* \leq 220$, $f_3^* \approx 2.5$, thus including values other than those found from Table 8-5; for example, $f_2^* = 210$, $f_3^* = 2.5$ is also a preferred solution.

Reversion to the decision space is performed for values in the indifference band, using the GRG program. Results for three equidistant values in the indifference band are presented in Table 8-5.

Again the surrogate worth trade-off method is able to develop a solution even though the multiobjective problem is highly non-linear with 20 decisions and 30 constraints and three objectives (of which two are of the minimax format). Although a real decision maker was not utilized, the SWT

method could have found the indifference band for any DM responses.   The
solution of the multiobjective problem in ε-constraint format is the most
difficult part of the procedure, but the availability of the GRG package
for solving non-linear problems rendered this step simple.   In addition,
the Lagrange multipliers for the ε-constraints are automatically output by
the GRG.

The actual decision-maker for this model of the Trinity River subsystem would of course select a different set of values of water power and
cost than was selected by the substitute decision-maker used for this example.   However, the number of iterations and the form of the calculation is
essentially the same.

It is, of course, possible to assign prices to the energy and water
produced as was done by Hall and Shephard[9]. This procedure in effect, pre-
sumes that all units of water and all units of energy are indistinguishable
in value regardless of cost of production or levels of production. In some
cases this is approximately correct, particularly when the power and water
contracts specify fixed prices and when the only interest of the decision
maker is the monetary return in excess of cost.   In most civil systems,
however, the latter is not strictly true, even when the prices producing
revenue are fixed in advance of analysis. The simplicity of the SWT method
is immediately apparent if the "value" of water and energy is not based on
an arbitrary price but rather on their impact on social goals.   Evaluating
a "price for all combinations of water, energy and cost levels for the 600
month planning horizon would be a difficult task.

In this particular example it is possible to estimate the "cash value"
of the marginal units of water and energy for the preferred solutions. Note
that the product of the true-worth ratio times the trade-off ratio must equal -1 at optimality.   Since the DM considers that the preferred solutions
he attained are optimal, λ^*_{12} is the marginal monetary price of MWH, i.e.,
$37 per megawatt-hour at the margin in our example, and $3.60 per ac-ft. is
the monetary value of water at the margin for our DM. These are the average unit
values and nothing can be deduced regarding the total value of the water
and power produced except that their sum exceeds the total cost of production.

TABLE 8-6

Preferred Solutions for California Water Project Example

i		1	2	3	4	5	6	7	8	9	10
$f_1^* = 95$	x_i	3.23	3.81	2.79	2.77	2.74	2.73	2.74	2.72	2.73	2.71
$f_2^* = 200$	r_i	.73	.31	.29	.27	.24	.23	.23	.22	.22	.21
$f_3^* = 2.5$	q_{i+1}	193.0	209.8	228.8	252.7	274.0	282.5	286.1	289.6	298.1	308.5
$f_1^* = 97$	x_i	3.29	2.86	2.84	2.82	2.78	2.79	2.77	2.76	2.76	2.77
$f_2^* = 210$	r_i	.79	.36	.34	.32	.29	.28	.27	.27	.26	.26
$f_3^* = 2.5$	q_{i+1}	193.0	209.7	228.7	252.5	273.7	282.3	285.8	289.2	297.7	308.0
$f_1^* = 99$	x_i	3.35	2.91	2.89	2.85	2.85	2.83	2.82	2.81	2.81	2.80
$f_2^* = 220$	r_i	.84	.41	.39	.35	.34	.33	.32	.32	.31	.30
$f_3^* = 2.5$	q_{i+1}	192.9	209.6	228.5	252.3	273.5	282.0	285.4	288.9	297.2	307.6

Note: x_i, r_i and q_{i+1} are all in units of KAF.

FOOTNOTES

1. This multiobjective problem is described in Reid and Vemuri [1971] and Vemuri [1974]; their solution was to find an analytic function for the non-inferior set.

2. Rows 16-24 in Table 8-1 correspond to the values taken from Reid-Vemuri [1971]. Note that their J_1^* corresponds to f_2, and J_2^* to f_1^* due to the way the objectives were defined.

3. Some of the authors who have studied this problem are Hass [1970], Haimes et al [1972], and Liebman and Lynn [1966].

4. The original formulation of this equation is found in Streeter and Phelps [1925].

5. A description of upper bounding techniques can be found in Taha [1971].

6. Both the data and the basic model are taken from Hass [1970].

7. See Hall and Shephard [1967] or Hall and Dracup [1970] for a description of the California water supply system and its analysis in the single objective format

8. This algorithm is described by Lasdon et al [1973].

9. Again see Hall and Shephard [1967].

REFERENCES

1. Haimes, Y. Y., Kaplan, M. A., and Husar, M.A., "A Multilevel Approach to Determining Optimal Taxation for the Abatement of Water Pollution," Water Resources Research, vol. 8, no. 4, 1972.

2. Hall, W. A. and Dracup, J. A., Water Resources Systems Engineering, McGraw-Hill Book Company, N.W., 1970.

3. Hall, W.A. and R.W. Shephard, "Optimum Operations for Planning of a Complex Water Resources System," University of California Water Resources Center. Contribution #122, Los Angeles, 1967.

4. Hass, J. E., "Optimal Taxing for the Abatement of Water Pollution," Water Resources Research, vol. 6, no. 2, 1970.

5. Lasdon, L. S., Fox, R. L., and Ratner, M W., "Nonlinear Optimization Using the Generalized Reduced Gradient Method," Case Western Reserve University, Operations Research Department Technical Memorandum no. 325, 1973.

6. Liebman, J.C., and Lynn, W.R., "The Optimal Allocation of Stream Dissolved Oxygen," Water Resources Research, vol. 2, no. 3, 1966.

7. Reid, R. W., and Vemuri, V., "On the Non-inferior Index Approach to Large Scale Multi-Criteria Systems," Journal of the Franklin Institute, vol. 291, no. 4, 1971.

8. Streeter, H.W., and Phelps, E.B., "Study of the Pollution and Natural Purification of the Ohio River," Public Health Bulletin no. 146, 1925.

9. Taha, H. A., Operations Research: An Introduction, The Macmillan Company, N. Y., 1971.

10. Vemuri, V., "Multiple Ojbective Optimization in Water Resource Systems," Water Resources Research, vol. 10, no. 1, 1974.

Chapter 9

MULTIOBJECTIVE WATER QUALITY MODELS

9.1   INTRODUCTION

In a world where new crises continue to overshadow previous ones and
the attempted solution of one crisis is certain to affect previous crises
and create new ones, where multiple and often noncommensurable goals and
objectives (often in conflict and competition with each other) characterize
our society, a cautious approach to systems modeling and optimization is
needed.   This sober and realistic approach should recognize the mutual interactions among the various goals and objectives and should aid the decision maker(s) in analyzing the trade-offs among various objectives in a
quantitative way.

A fundamental and almost axiomatic prerequisite for these models to
be realistic, and thus be considered for an ultimate utilization by the decision maker(s), is that they be susceptible to multiple objective functions
in their noncommensurable forms and units.

The second fundamental prerequisite is that there should exist solution methodologies which are capable of analyzing and optimizing (in the
mind of the decision maker(s)) these multiple objectives.

The present lack of mathematical models with multiple objective functions can be attributed primarily to the past lack of operational methodologies capable of analyzing and optimizing multiple noncommensurable objective functions.   The Surrogate Worth Trade-off (SWT) Method is one such
methodology that fulfills the second prerequisite.   Systems engineers and
systems modelers can now move to fulfill the first prerequisite for realistic models by constructing the proper multiple objective functions.

9.2   WATER QUALITY GOALS AND OBJECTIVES

Two major classes of water quality objectives may be identified. The
first class, primary objectives, corresponds directly to water quality standards and thus depends upon the particular water resource's physical, chemi-
cal, and biological characteristics.

Secondary objectives, the second classification, correspond both to
the impact of water quality level on the utilization of the water resource
and to the impact of water resource use on quality.

9.2.1 Primary Objectives

The Great Lakes Water Quality Agreement between the U.S. and Canada
provides a listing of primary objectives and goals in water quality[1]. These

objectives and goals are presented and specified in the Act either in terms
of upper or lower constraints (levels of achievement) or in terms of objectives to be minimized or maximized. The following is a sample of water
quality goals and objectives:

minimize { Phenols and other objectionable taste and odor
          { producing substances

minimize { Temperature change that would adversely affect any
          { local or general use of the water

minimize { Mercury and other toxic heavy metals
          { Persistent pest control products and other

minimize { persistent organic contaminants that are toxic or
          { harmful to human, animal, or aquatic life

minimize { Settleable and suspended materials}

minimize { Oil, petrochemicals, and immiscible substances}

minimize { Radioactivity}

and goals in terms of upper and lower constraints:

total coliforms	$\leq$ 1000 per 100 milliliters
fecal coliforms	$\leq$ 200 per 100 milliliters
dissolved oxygen	$\leq$ 6.0 milligrams per liter
total dissolved solids	$\leq$ 200 milligrams per liter
Iron	$\leq$ 0.3 milligrams per liter
6.7 $\leq$ pH	$\leq$ 8.5

In most cases, the upper or lower achievement actually corresponds
to ultimate desired goals representing short, intermediate, or long term
perspectives. Quantitative consideration of these goals as constraints
within a formal model may, therefore, introduce a severe intractability in
their further analysis, as well as a misrepresentation of the real world
that is being modeled.

Two conclusions that may be drawn from the above goals and objectives:

(i) There exist multiple noncommensurable objectives in water
     quality control and management.

(ii) Objectives and goals included in a mathematical model as
     constraints (such as upper limit on coliforms or lower limit
     on dissolved oxygen) will generally (and most likely) lead
     to solutions on the boundaries of those constraints. Thus
     important information concerning the effect of relaxing one
     constraint on the improvement of another is not readily
     available.

9.2.2 Secondary Objectives

In general, the secondary objectives are not defined as precisely or quantitatively as the primary objectives. The following is a sample of secondary objectives:

(i)   Reduction in the level of algal growth

(ii) Restoration of year-round aerobic conditions

(iii) Restoration of the water body for the purpose of swimming, fishing, and recreation

(iv) Minimization of any health hazards

Clearly, a systematic and quantitative methodology capable of analyzing the trade-offs among all objectives is needed.   The Surrogate Worth Trade-off (SWT) Method fills this need as will be discussed subsequently.

9.3   GENERAL PROBLEM FORMULATION

Most existing single objective models for water quality control and management can be extended to include multiple objective functions.   As an illustration, a single objective function water quality model[2] will be extended to the case of multiple objectives.

Given a water resources system, it is convenient to decompose the system into N subsystems (N reaches in the case of a river). This allows modeling and analysis of all of the system inputs and responses.

Let the vector $\underline{U}_i$ be the input (pollution) to the ith subsystem, $i = 1,2,\ldots,N$, where $\underline{U}_i = [U_{i1}, U_{i2},\ldots,U_{iM}]$. The first element of the vector $\underline{U}_i$, U_{i1}, may represent water quantity and the $U_{i2},\ldots,U_{iM}$ may represent different water quality characteristics (e.g., BOD, pH, temperature, total dissolved solids, etc.).

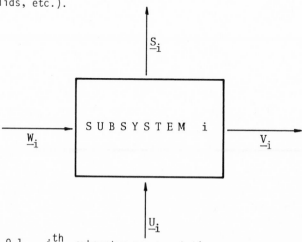

Figure 9.1 i^{th} subsystem representation

Let

$\underline{W}_i$ be the input vector coming into the ith subsystem from other subsystems,

$\underline{V}_i$ be the output vector of the ith subsystem going to other subsystems, and

$\underline{S}_i$ be the decision vector of the ith subsystem,

where $\underline{W}_i$, $\underline{V}_i$, and $\underline{S}_i$ are of the same dimension as $\underline{U}_i$, i.e., M-dimension (see Figure 1). Clearly, the vector sum $\underline{W}_i + \underline{U}_i$ is meaningless and does not equal $\underline{V}_i + \underline{S}_i$. The subsystem outputs $\underline{V}_i$ and $\underline{S}_i$ are assumed to be represented by the following functions.

$$\underline{V}_i = \underline{\Psi}_i \ (\underline{U}_i, \ \underline{W}_i)$$

$$\underline{S}_i = \underline{\Phi}_i \ (\underline{U}_i, \ \underline{W}_i)$$

where

$$\underline{\Psi}_i = [\Psi_{i1}, \ \Psi_{i2}, \ \ldots, \ \Psi_{iM}]$$

$$\underline{\Phi}_i = [\Phi_{i1}, \ \Phi_{i2}, \ \ldots, \ \Phi_{iM}]$$

$$i = 1, 2, \ldots, N$$

At present, not all components of the functions $\underline{\Psi}_i$ and $\underline{\Phi}_i$ are known to water quality experts. This however constitutes no limitation to the model, because whenever the functional relationship of any component of $\underline{U}_i$ is known, that component can be inserted and thus be considered in the analysis. Naturally, the presence or absence of a component of $\underline{U}_i$ has a strong effect on the model. Many current mathematical models consider as components of $\underline{U}_i$ only BOD or DO and assume the Streeter-Phelps functional relationship[3]. It is evident, however, that the greater the number of components of $\underline{U}_i$ that are considered and analyzed, the more accurate and representative the mathematical model becomes.

Finally, let

$$\underline{G}(\underline{U}, \ \underline{W}, \ \underline{S}) \leqslant \underline{0}$$

be k-dimensional vector of constraints. The inequality constraints include equality constraints and represent the physical, legal, economic, and other system constraints.

Note that

$$\underline{U} = [\underline{U}_1 : \underline{U}_2 : \quad : \underline{U}_M]$$

$$\underline{S} = [\underline{S}_1 : \underline{S}_2 : \quad : \underline{S}_M]$$

$$\underline{W} = [\underline{W}_1 : \underline{W}_2 : \quad : \underline{W}_M]$$

are augmented vectors.

The single criteria optimization problem, e.g., the minimization of the sum F^1 of treatment cost functions of each subsystem, $F_i(\underline{U}_i, \underline{S}_i)$:

$$F^1(\underline{U},\underline{S}) = \sum_{i=1}^{N} F_i(\underline{U}_i, \underline{S}_i) \tag{1}$$

can be extended to include other objectives. For example, let

$$F^j(\underline{U},\underline{S}) = \sum_{i=1}^{N} F_i^j(\underline{U}_i,\underline{S}_i), \quad j = 1,2, \ldots, J$$

be the jth objective function for the water resources system, e.g.:

for $j = 1$, F^1 is the above cost function, (Eq. 1);

for $j = 2$, F^2 may be total phenols;

for $j = 3$, F^3 may be total mercury and other toxic heavy metals;

for $j = 4$, F^4 may be total settleable and suspended materials;

for $j = 5$, F^5 may be total oil, petrochemicals, and immiscible substances.

Clearly, the overall objective function for the water resources system will generally not be a simple summation of the various objectives. Also, note that there is no discrepancy in the fact that the mth component of $\underline{U}_i$, U_{im}, may be mercury concentration and that the jth objective function F^j is also mercury concentration.

The overall model for water quality control and management can be written as:

System 9-1:

minimize $\{F^1(\underline{U}, \underline{S}), \ldots, F^J(\underline{U}, \underline{S})\}$
$\underline{U},\underline{S}$

subject to the constraints

$\underline{G}(\underline{U}, \underline{W}, \underline{S}) \leq \underline{0}$

Decomposition and Multilevel Approach[4] can be applied for solving the overall problem where the Surrogate Worth Trade-off Method is utilized as a higher level coordinator. The discussion of the hierarchical-multilevel approach is beyond the scope of this book.

In the following section, examples for multiobjective systems will be discussed.

9.4   FORMULATION OF AN EXAMPLE PROBLEM

The following formulation is an integration and extension of several single objective function models.[5]

In the model discussed here, the vectors $\underline{U}_i$, $\underline{S}_i$, $\underline{W}_i$, and $\underline{V}_i$ (for the ith subsystem) are three dimensional vectors; the elements of which are associated with biological oxygen demand load, thermal load, and algae concentration respectively.

Models which are concerned with the effect of biological oxygen demand (BOD) load on the dissolved oxygen (DO) in the stream are similar to the stream resource allocation model described in Chapter 8. The single objective function is the total cost of wastewater treatment for all users along the stream.

The stream is segmented into N reaches, each of which is associated with users (polluters) who may discharge organic wastes into the stream.

The Streeter-Phelps relation[6] is utilized to transform minimum dissolved oxygen standards for each reach into a set of linear inequalities relating upstream treatment levels to downstream decisions.

The constraints require that the supply of available oxygen for the organic decomposition process in each reach (that available above the quality standard requirement) must be equal to or exceed the demand imposed by BOD loads discharged into that reach and all reaches preceding it.

Thus, for reach i it is required that

$$d_{i1} u_{11}(1 - s_{11}) + d_{i2} u_{21}(1 - s_{21}) + \ldots$$

$$+ d_{ii} u_{i1}(1 - s_{i1}) \leq e_i \qquad i = 1, 2, \ldots, N \qquad (2)$$

where u_{j1} = gross biological oxygen demand (BOD) load introduced at the beginning of the jth reach that has a polluting input (lbs/day).

s_{j1} = fraction of u_{j1} removed through treatment by the jth polluter.

d_{ij} = pounds of oxygen demanded by the decomposition of a pound of BOD discharged by the jth polluter in reach i.

e_i = amount of dissolved oxygen available for the decomposition process (total available less standard requirement) in reach i per unit of flow.

In addition, other restrictions on s_{j1}'s require at least primary treatment, thus

$$0.45 \leq s_{j1} \leq 0.99 \qquad j = 1, 2, \ldots, N$$

An objective function associated with this model is to minimize the total cost of wastewater treatment. $\bar{F}^1(\underline{S}_1)$:

$$\underset{s_{j1}}{\text{Min}} \quad \bar{F}^1(\underline{S}_1)$$

where

$$\bar{F}^1(\underline{S}_1) = \sum_{j=1}^{N} \bar{f}_j^1(s_{j1})$$

$$\underline{S}_1 = [s_{11}, s_{21}, \ldots, s_{N1}]$$

and $\bar{f}^1(s_{j1}) = 160.8 + 26.7\, q_j + 640.7\, (s_{j1} - 0.45)^2$

$$+ 255.7\, q_j\, (s_{j1} - 0.45)^2 \tag{3}$$

where $0.45 \leq s_{j1} \leq 0.99$

and, q_j is the plant size in million gallons per day. Hass derived equation (3) from Frankel's data for the Miami River in Ohio.[7]

The model presented by Foley, and Foley and Haimes[8] concerned itself with other water quality standards in addition to DO. These were thermal pollution and algae.

Let:

u_{j2} = raw load of energy introduced at the beginning of the jth reach

s_{j2} = percentage of waste heat (u_{j2}) removed by cooling towers

v_{j2} = temperature of water leaving the jth reach to the $j + 1^{st}$ reach

A set of constraints on the thermal pollution can be introduced.

$$v_{j2} \leq v_{j2}\,(\text{Max})$$

$$v_{j2} = v_{j2}(u_{j2}, s_{j-1,2}, s_{j-2,2}, \ldots, s_{12}, v_{02}, t) \tag{4}$$

where

$v_{j2}(\text{Max})$ = maximum temperature allowed

v_{02} = initial temperature of flow entering the first reach

t = time

Equation (4) can be rewritten as:

$$v_{j2} = v_{j2}(w_{j2}, u_{j2}, t)$$

where

w_{j2} = temperature of the water entering the jth reach from the (j-1) reach at time t.

Critical flow conditions imply dropping the time dependence in the exponential decay as given by Lesbosquet[9]. In a development similar to equation 2, the decisions s_{j2} are related to the temperature v_{j2} by the following equation:

$$v_{j2} = \sum_{k=1}^{j} b_{kj} \, u_{k2} \, (1-s_{k2}) \leq v_{j2} \, (\text{Max}), \qquad j = 1,2, \ldots, N \qquad (5)$$

where b_{kj} = constants relating the treatment of the raw load u_{k2} to a decrease in temperature in the jth reach.

Note that the u_{j2} are given in units commensurable with v_{j2} by assumption, and that the entire flow of the river is utilized by the stream power plants; otherwise equation (5) must be modified.

The cost function must be modified in the present context to include the additional cost associated with the removal of thermal pollution, $\hat{F}^1(\underline{S}_2)$.

$$\hat{F}^1(\underline{S}_2) = \sum_{j=1}^{N} \hat{f}_j^1 \, (s_{j2})$$

where $\hat{f}_j^1(s_{j2}) = \alpha_j \, s_{j2}^2 + c_j$,

$$\underline{S}_2 = [s_{12}, \, s_{22}, \, \ldots, \, s_{N2}]$$

and α_j and c_j are cost coefficients of the system. Since it is desired to minimize the cost of treatment of thermal pollution, $\hat{F}^1(\underline{S}_2)$ can be added to the cost function. This yields a new cost function

$$F^1(\underline{S}_1, \, \underline{S}_2) = \{ \bar{F}^1(\underline{S}_1) + \hat{F}^1(\underline{S}_2) \}$$

Therefore, the first objective function (in commensurable dollars) is:

$$\underset{\underline{S}_1, \underline{S}_2}{\text{Min}} \quad F^1(\underline{S}_1, \, \underline{S}_2)$$

subject to the constraints discussed previously. The objective function for thermal pollution expresses the desire to minimize the temperature change that would adversely affect any local or general use of the water. Thus,

$$\underset{\underline{S}_2}{\text{MIN}} \quad F^2(\underline{V}_2, \, \underline{S}_2) = \sum_{j=1}^{n} (v_{j2} - v_{j-1,2})^2$$

Since $\underline{V}_2$ is given as a function of $\underline{S}_2$ (in equation 5) the second objective function can be rewritten as a function of $\underline{S}_2$ only:

$$\text{MIN}_{\underline{S}_2} \; F^2(\underline{S}_2) \; = \; \sum_{j=1}^{n} \; (b_{jj} u_{j2} \, [1 - s_{j2}])^2$$

The objective for algae concentration is derived as follows:
Following Bailey[10]

$$v_{j3} \; = \; \alpha_1 \, \exp \, \alpha_2 \, (\alpha_3 + 0.16 \, v_{j2}) \hspace{2cm} j = 1,2, \ldots, N \quad (6)$$

where

v_{j3} = algae (phytoplankton) concentration at the end of the jth reach
and α_k, $k = 1,2,3$ are constants characteristic to the stream. In a preliminary study, these constants are determined by the assumed critical
values of water depth, solar intensity, and nutrient concentration. Note
that algae growth depends on thermal load removal $\underline{S}_2$ via the v_{j2} in equation (6).

The objective function represents the desire to minimize the maximum
algae concentration for all reaches. Thus

$$F^3(\underline{S}_2) \; = \; \text{Max}_j \; v_{j3}, \hspace{1cm} j = 1,2, \ldots, N$$

To summarize, the overall mathematical model includes three noncommensurable objective functions, as well as several constraints:

$$\text{MIN} \; F^1(\underline{S}_1, \underline{S}_2) \; = \; \sum_{j=1}^{n} \; 160.8 + 26.7 \; q_j + 640.7 \; (s_{j1} - .45)^2$$

$$+ \; 255.7 \; q_j \; (s_{j1} - .45)^2 + a_j s_{j2}^2 + c_j$$

$$\text{MIN} \; F^2(\underline{S}_2) \; = \; \sum_{j=1}^{n} \; (b_{jj} \, u_{j2} \, [1 - s_{j2}])^2$$

$$\text{MIN} \; F^3 \, (\underline{S}_2) \; = \; \text{Max}_j \; v_{j3}$$

Subject to

$$.45 \; \leq \; s_{j1} \; \leq \; .99 \hspace{1cm} j = 1,2, \ldots, N$$

$$0 \; \leq \; s_{j2} \; \leq \; 1.00 \hspace{1cm} j = 1,2, \ldots, N$$

$$\sum_{k=1}^{j} \; d_{jk} u_{k1} (1 - s_{k1}) \leq e_i \hspace{0.7cm} j = 1,2, \ldots, N$$

$$\sum_{k=1}^{j} \; b_{kj} \, u_{k2} \, (1 - s_{k2}) \leq v_{j2 \, \text{Max}} \hspace{0.5cm} j = 1,2, \ldots, N$$

9.5 APPLICATION OF THE SWT METHOD TO THE THREE WATER QUALITY OBJECTIVE PROBLEM

The three objectives defined in the previous section are to be optimized as symbolically expressed by system 9-1, subject to the existing physical constraints.

In order to generate the trade-off functions, λ_{ij}, the vector optimization problem is rewritten in the ε-constraint form as follows:

$$\min_{\underline{S}_1, \underline{S}_2} \quad F^1(\underline{S}_1, \underline{S}_2)$$

subject to the constraints

$$F^2(\underline{S}_2) \leq \varepsilon_2$$
$$F^3(\underline{S}_2) \leq \varepsilon_3$$
$$\underline{G}(\underline{U}, \underline{W}, \underline{S}) \leq \underline{0}$$

where the variables ε_j are related to $\bar{f}_j$ (the minimum of the jth objective function, while ignoring all other $(n-1)$ objectives) as follows:

$$\bar{f}_j = \min F^j$$
$$\varepsilon_j = \bar{f}_j + \delta_j$$

where $\delta_j > 0$

The system's Lagrangian, L, is:

$$L = F^1 + \lambda_{12}(F^2 - \varepsilon_2) + \lambda_{13}(F^3 - \varepsilon_3) + \underline{\mu} \cdot \underline{G} \qquad (7)$$

where $\underline{\mu}$ is a vector of Lagrange multipliers.

The trade-off functions λ_{12} and λ_{13} are determined by solving equation (7). A detailed computational discussion on the construction of the trade-off functions was given in Chapter 3.

Note that the value of λ_{12} is the ratio of the incremental gain in objective 1 (cost minimization) to the incremental loss in objective 2 (temperature) and a value of λ_{13} is the trade-off ratio between cost and algae production. The λ_{ij} corresponding to the binding constraints are associated with the noninferior solution and thus are of interest, these λ_{ij} are also positive. The term $(F^2-\varepsilon_2)$ represents the amount by which F^2 exceeds the target attained level of temperature, ε_2. The Lagrange multiplier λ_{12} which makes $\lambda_{12}(F^2-\varepsilon_2) = 0$ when equation (7) is minimized is the "shadow price" or marginal trade-off value, expressed in dollar cost per unit temperature increase.

The Lagrangian is solved for different values of ε_2 and ε_3. Corresponding to each solution are a minimum of objective function 1, F^{1*}, and values of ε_2, ε_3, λ_{12}, and λ_{13}. By changing ε_2 and ε_3 over a reasonable range, the trade-off rate functions $\lambda_{12}(F^{1*},\varepsilon_2,\varepsilon_3)$ and $\lambda_{13}(F^{1*},\varepsilon_2,\varepsilon_3)$ can be computed.

Rewriting the vector minimization problem posed in the ε-constraint form where F^2 is the primary objective yields the same solution. The trade-off rate functions λ_{21} and λ_{23} corresponding to the Lagrangian, L_2,

(where $L_2 = F^2 + \lambda_{21}(F^1 - \varepsilon_1) + \lambda_{23}(F^3 - \varepsilon_3) + \underline{\mu} \cdot \underline{G}$) ,

are related to λ_{12} and λ_{13} as described in Chapter 6.

The above steps can be taken on a strictly analytical basis without an interaction with the decision-maker. The following steps involve the decision-maker.

Select any set (F^{1*}, ε_2, ε_3) as "optimized" attained levels of cost, temperature rise, and algae production resulting from the "optimal" decision determined by solving equation (7). For this set λ_{12} and λ_{13} are known. Begin with λ_{12}. The decision-maker is asked whether or not he would give up one unit of temperature in order to gain λ_{12} units of cost. If he says yes, he is asked to assign a numerical value between 0 and +10 to show how strongly he would feel about that trade, zero being indifferent, 10 representing a strong drive to gain cost at the expense of temperature. If he says no, then he assigns a value between 0 and -10 to represent how strongly he feels toward the opposite direction of trading. The surrogate worth value corresponding to ε_2 and ε_3 (assigned by the decision-maker) is denoted by $W_{12}(\varepsilon_2, \varepsilon_3)$.

At the same time the decision-maker is asked his preference with respect to λ_{13} in exactly the same way. He is, of course, informed what the attained levels of cost, temperature, and algae production would be (F^{1*}, ε_2, ε_3). Note that $F^{1*} = F^{1*}(\varepsilon_2,\varepsilon_3)$ hence there are really only two independent variables in the objective space.

Let us presume, for the sake of illustration, that the decision-maker gave a value of +8 to the surrogate worth of λ_{12} and a value of +3 to the surrogate worth of λ_{12} at other ε_2, ε_3. This indicates a decrease in both ε_2 and ε_3 are required. The decision-maker is then asked for the surrogate worth value, $W'(\varepsilon_2',\varepsilon_3')$, corresponding to $F^{1*}(\varepsilon_2', \varepsilon_3')$, λ_{12}', and λ_{13}', where $\varepsilon_2' < \varepsilon_2$, $\varepsilon_3' < \varepsilon_3$. The decision-maker is asked to make a "consistent" estimate of $W'_{12}(\varepsilon'_2, \varepsilon'_3)$ and $W'_{13}(\varepsilon'_2, \varepsilon'_3)$. That is, $W'_{12} < W_{12}$ and $W'_{13} < W_{13}$. Since his past value of W_{12} was +8 he might say +4 if he felt you had made a considerable improvement but he would still trade temperature for cost reduction rather emphatically. He might say +1 if he wasn't quite so emphatic. Similar analysis can be performed for W'_{13}.

With these two sets of values of the surrogate worth functions W_{12} and W_{13} it is now possible to make a linear interpolation (or extrapolation) to find the point $(\varepsilon^o_2, \varepsilon^o_3)$ at which both W_{12} and W_{13} would equal zero if the surrogate worth functions were linear. This point, $(\varepsilon^o_2, \varepsilon^o_3)$, is then used as a third trial value and the process repeated until the decision-maker is unable to say with certainty whether he would trade further or not. Such a situation corresponds to a zero of the surrogate worth function and to the value of the real worth function which equates the worth of the gain in cost reduction to the worth of the loss in temperature control and that of cost reduction to algae control. This is a preferred solution in the sense that no knowledge exists by which the decision-maker could assert a "better" solution.

By carefully approaching the zero of the surrogate worth function from the positive side only (i.e., + values) a "left hand" bound on the band of indifference can be found. By repeating the process from the negative side a "right hand" bound can be determined. If both bounds are at the same value of $(\varepsilon_2, \varepsilon_3)$ for both W_{12} and W_{13}, then a unique solution has been found. If not, then the range $\varepsilon_{2\ell} \leq \varepsilon_2 \leq \varepsilon_{2r}$ and $\varepsilon_{3\ell} \leq \varepsilon_3 \leq \varepsilon_{3r}$ is the band of indifference and any $(\varepsilon_2, \varepsilon_3)$ in this range is as good as any other.

All decisions, $\underline{S}_1$ and $\underline{S}_2$, pertaining to any preferred solution are also implied for the band of indifference and can be directly calculated as described in Chapter 6.

9.6 SUMMARY AND CONCLUSIONS

The Surrogate Worth Trade-off Method is particularly useful for water quality problems where cost, dissolved oxygen, temperature, BOD, etc. are the measures of goals, but for which no rational procedure for commensuration in common units is available or likely to become available. The "worths" of these levels of goal attainment are not universal but are very site sensitive. This procedure avoids the problem of attempting to determine common utility so that optimization can be accomplished and, instead, accomplishes a functional optimization in multiple objective space and then evaluates only parity of trade-off in this optimized objective space.

For the procedure to be strictly correct the measures of the objectives (DO, pH, etc.) must be either true measures of the actual objectives or be monotonically related to them. A good example of a measure which does not meet this criteria is the use of "visitor-days" as a measure of an objective to maximize a recreational objective. Obviously, as this index increases the objective is enhanced, but only up to some unknown limit or saturation point beyond which additional visitor days can turn an expected recreational experience into a nightmare. However, such "improper" indices will reveal themselves by anomalies in decision-maker responses so that no practical harm is done.

FOOTNOTES

1.   The entire list can be found in Great Lakes Water Quality [1972].

2.   This model is described in detail by Haimes [1971].

3.   The derivation of this relationship can be found in Streeter and Phelps [1925].

4.   The multilevel approach is presented in Haimes [1973].

5.   The following multiobjective model was presented by Haimes and Hall [1975]. Among the single objective water quality models are those of Haas [1970], Haimes, Foley and Yu [1972], Haimes, Kaplan and Husar [1972], Foley [1971] and Foley and Haimes [1973].

6.   Again, see Streeter and Phelps [1925].

7.   See Haas [1970] for the derivation and Frankel [1965] for the data.

8.   See Foley [1971] and Foley and Haimes [1973].

9.   This relationship was originally derived by Lesbosquet [1946].

10.   The derivation of this equation is presented by Bailey [1970].

References

1.   Bailey, T.E., "Estuarine Oxygen Resources-Phytosynthesis and Reaeration", Journal of the Sanitary Engineering Division, ASCE, vol. 96, no. SA2, Proc. Paper 7215, April 1970, pp. 279-296.

2.   Foley, J. W., Multilevel Control of Water Quality, M.S. Thesis, Case Western Reserve University, Cleveland, Ohio, June 1971.

3.   Foley, J. W., and Y. Y. Haimes, "Multi-level Control of Multi-pollutant System", ASCE Journal of the Environmental Engineering Division vol. 99, no. EE3, pp. 253-268, June 1973.

4.   Frankel, R. J., "Water Quality Management: An Engineering-Economic Model for Domestic Waste Disposal, Ph.D. Dissertation, University of California, at Berkeley, Calif., 1965.

5.   Great Lakes Water Quality, agreement between the United States of America and Canada, signed at Ottawa, April 15, 1972.

6.   Haimes, Y. Y., and W. A. Hall, "Multiobjectives in Water Resources
      Systems Analysis: The Surrogate Worth Trade-off Method",   Water
      Resources Research, vol. 10, no. 4, August 1974, pp. 615-624.

7.   Haimes, Y.Y., "Modeling and Control of the Pollution of Water Resour-
      ces Systems Via Multilevel Approach", Water Resources Bulletin, vol.
      7, no. 1, Feb. 1971, pp. 104-112.

8.   Haimes, Y.Y., and Hall, W. A., "Analysis of Multiple Objectives in
      Water Quality", Presented at the special ASCE Conference at Cornell
      University, Ithaca, N.Y., June 26-28, 1974. To appear in the Jour-
      nal of ASCE Hydraulic Division, 1975.

9.   Haimes, Y.Y., "Decomposition and Multilevel Approach in Modeling and
      Management of Water Resources System", pp. 348-368,   Decomposition
      of Large Scale Problems, D. M. Himmelblau, Editor, North Holland
      Publishing Co., Amsterdam, 1973.

10.   Haimes, Y. Y., Foley, J. W., and Yu, W.,"Computational Results for
      Water Pollution Taxation Using Multilevel Approach," Water Resources
      Bulletin, vol. 8, no. 4, Aug. 1972, pp. 761-772.

11.   Haimes, Y.Y., Kaplan, M. A., and Husar, M. A., "A Multilevel Approach
      to Determining Optimal Taxation for the Abatement of Water Pollu-
      tion", Water Resources Research, vol. 8, no. 4, Aug. 1972, pp. 851-
      860.

12.   Haas, J. E., "Optimal Taxing for the Abatement of Water Pollution,"
      Water Resources Research, vol. 6, no. 2, April 1970, pp. 353-365.

13.   Lesbosquet, M., "Cooling-Water Benefits from Increased River Flows,"
      Journal of the New England Water Works Association, vol. 60, June
      1946, pp. 111-116.

14.   Streeter, H.W., and Phelps, E. B., "Study of the Pollution and Natu-
      ral Purification of the Ohio River", Public Health Bulletin No.
      February 1925.

Chapter 10

SENSITIVITY, STABILITY, RISK AND IRREVERSIBILITY

AS MULTIPLE OBJECTIVES

10.1 INTRODUCTION

Water resources projects are planned, designed, constructed, opera-
ted and modified under numerous risks and uncontrollable uncertainties.
While in general the terms risk and uncertainty can denote the same thing
(it is risky because it is uncertain) it is useful for analytical purposes
to define these separately as two distinct concepts. Risk is characterized
by a frequency distribution of events following reasonably well known or
measurable probabilities, even though the specific time or spatial sequence
of occurence of events cannot be determined. In water resources problems
for example, a common cause of risk is the associated hydrologic input.
Past hydrological records are usually available to reasonably define the
probability distribution, but any specific sequence of events is largely
controlled by chance. In contrast to risk, uncertainty is characterized by
the absence of any known reasonably valid probability distribution of
events. The term risk is assigned to measurable chance controlled factors,
while uncertainty applies to all others. In water resources for example,
there are uncertainties associated with:

- The growth of population, industry, agriculture and
  urban areas.
- The projected cost of labor, material, and inflation.
- The assessment of future advancement in engineering,
  science, and technology.
- The projected benefits associated with the projects.

In addition, there are important uncertainties and risks introduced by both
the system and man's attempts to model it.

There are many types of risk and uncertainty in water resources,
most of them well known to the practitioners in the field. Of these, the
one most important for the decision process modeler is that related to the
precision with which the control variables can in fact be controlled. An
"optimal" solution in the sense of minimizing the distance, time or cost
involved in crossing a deep gorge might be to cross hand and hand on any
convenient cable such as a high voltage electrical conductor. For most of
us, however, such a decision would be unthinkable because we know that we
lack the necessary precision of control with respect to this particular de-

cision variable. While an extreme example, it does serve to illustrate
that the classical concept of optimum (signifying best) is by itself inade-
quate -- the degree of control of the significant system responses must be
also considered as a decision parameter for a great many water resource
systems.

To the extent that the effect of the lack of control can be characterized by a probability distribution, the corresponding uncertainty is reduced to risk, and in a few cases in water resources systems, but by no
means all (or even most), risk can to some extent be allowed for by optimizing the mathematical expectation of the results. In fact, where mathematical expectation is a valid criteria for optimality, it can be shown that it
is not always necessary to know the probability distribution in any great
detail.

If, on the other hand, the risk problem is characterized by infre-
quent decision, irreversibility, or both, optimization of mathematical expectation may lead to very serious errors. Since these two characteristics
tend to dominate the risk and uncertainty situations involved in water resources management, their proper consideration as separate noncommensurable
objectives is essential in most water resources decision models.

Other systems characteristics which may increase or mitigate the effects of risk and uncertainty are sensitivity, responsivity and stability.
Modeling inaccuracies can be important also. The discussion in this chapter of these system and modeling characteristics is intended to stimulate
the considerable amount of further research needed to evaluate their impact
on risk and uncertainty as objective functions. Some proposals on how this
might be done for the sensitivity characteristic are suggested. A better
knowledge and understanding of these characteristics and their relationship
to risk and uncertainty will permit these crucial factors to be properly
incorporated into multiobjective analysis.

10.2 SYSTEM CHARACTERISTICS RELATED TO THE EVALUATION OF RISK

Water resource systems have a number of characteristics associated
with the stochastic nature of the system inputs, outputs, and states. Four
of these characteristics are discussed here:

 · Sensitivity.
 · Responsivity
 · Stability.
 · Irreversibility.

These are the major elements of prototype systems involved in the definition of risks as indices of performance. As an additional characteristic,

the precision of the model representing the system will be discussed separately.   Although it is recognized that the current state-of-the-art in
systems analysis is not yet fully capable of quantitatively treating all of
these characteristics, it is essential that they be considered as thorough-
ly as possible. They are descriptively defined as follows:

Sensitivity is the system characteristic relating the changes in the
system's index of performance (or output) to expected variations in the decision variables, uncontrolled parameters, constraint levels or the model's
coefficients.

Responsivity is the system characteristic of being dynamically res-
ponsive to changes (including random variations) in the decisions over time.
This measures the ability of the significant responses of the system to
follow the changes in a variable decision in time and/or space.

Stability is a system characteristic related to the degree of variation of the mean system response to fixed decisions.   A stable system
yields an invariant mean response to the mean value of a decision set. A
system may be stable and still have an important random component.

Irreversibility is a system characteristic related to the degree of
difficulty involved in restoring previous states or conditions once the system has been altered by a decision (including the "decision" to do nothing).

Some examples of each of the above four characteristics will be
given in order to clarify the concepts.

10.2.1 Sensitivity

It is possible to construct hypothetical situations in which the deterministic mathematical optimum decision would be the worst possible unless the decision variable can be very precisely controlled.   Figure 10.1
illustrates such a situation in which it is presumed that the decision
variable can be controlled only within limits, x_c, and that x may take on
any value with equal likelihood within these limits.   The deterministic
mathematical maximum is obviously far from being the practical optimum decision. In this contrived example, x_2^* is clearly a "better" decision than
x_1^* unless the decision maker is more interested in gambling than risk avoidance.

Even if the example is treated by maximizing the mathematical expectation of f(x), it does not follow that a resulting "optimum" at x_1 is superior to x_2. For this to be true the appropriate objective must indeed be
the maximization or minimization of the expected value of f(x).   This is
seldom true where risk is a major consideration.   The "gamblers' ruin"
problem is the classical example where this is clearly not the objective.[1]

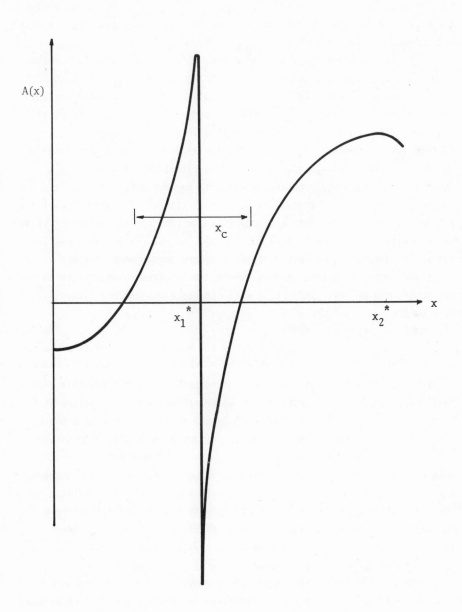

Figure 10-1. Sensitivity Band

As in many practical problems in water resources, the decision which maximizes the expected value of the return in this problem will also correspond
to a maximization of the risk of getting little or nothing.   In reality
there are two noncommensurable objectives in this case, avoidance of risk
and gaining economic return.

The presumption that the objective is the maximization of expected
value is simply one particular method of commensurating risk and return. It
would appear to be valid in those instances where the number of decisions to
be made over time and space is large enough to insure a high probability
that such a return will in fact be realized in the long run, despite intermediate ups and downs.   The expected value as an index of performance (objective function) clearly does not apply to situations in which only a very
few decisions (and their consequences) will be involved.   Nor is it valid
in any situation where the objective function itself is automatically and
discontinuously altered into a completely new form for all values of x, in
the event that x chanced to fall in a particular sub-range. There are large
numbers of situations in water resources where one or both of these excep-
tions are applicable and,   hence, where the expected value criterion for
commensuration of risk and return will not represent the objectives of the
real system.

10.2.2 Responsivity

Responsivity is the capability of the system to respond in a reasonable time frame to a variable (changing) decision. It it generally related
to "frictions" in the system and delayed response.   One of the most important responsivity characteristics of water and other civil systems is the
long lead time usually required to observe a need, to conceive a possible
means of meeting that need, to develop a preliminary plan, to obtain a
basic policital approval of the plan during a "political hassle period", to
complete the final design, and to construct or otherwise implement the decision. This is often in excess of twenty-years, and sometimes more than
forty or fifty years.   Even for small, almost inconsequential problems it
is seldom less than two years.   Since objectives can and do change at a
much more rapid rate, this form of responsivity has become exceedingly important in water resources planning.

There are many other forms of responsivity occurring in water resources.   A classic example is time delay in routing water down an open
channel aqueduct system.   Another is the related problem of flood routing.
Yet another is the ability of a "move" type supplemental irrigation system
to cover the entire field in the face of drought.   The response of hydro-

electric systems to rapid fluctuations in demand is an economically useful
responsivity of these systems.

The response of water use to price is another very important element
of responsivity in water resources systems.   In many instances involving
the use of water, costs which vary with the amount of water used are quite
small relative to costs which are insensitive to the volume of use (largely
irreversible capital investment).   This may result in a response delay of
the order of magnitude of the economic life of the investments involved.

10.2.3   Stability

Stability is a measure of the resistance to non-decision modification
of the mean response of the system. Frequently in water resources the response of the system will vary appreciably even for a fixed decision.   If
the effect of the variation is to return the system automatically to the
"output" or objective value represented by the decision, the decision system is stable. If, on the other hand, auto-catalytic effects cause the response to move away from that intended by the decisions, the decision system is unstable.   Many water resources and other civil systems have highly
unstable decision systems.   One obvious example is the flood control decision system.   It has been asserted that providing partial flood control,
commensurate with one set of predicted future conditions, has results in
attracting more economic activity into the "protected" area -- making the
original decision for partial control quite improper for the autocatalyti-
cally changed situation. Transportation routing is another classic example.
In most instances water based recreation has also responded in an unstable
way.   On the other hand, many of the past estimates of future water needs,
made up to fifty years in advance (e.g. Mulholland's almost perfect time-
frame estimate of need for 1600 cfs for Southern California coastal area)[2]
have proved to be remarkably uncanny, suggesting that highly stable self-
fulfilling effects may be involved in these cases.

10.2.4   Irreversibility

Irreversibility is a measure of the difficulty in returning a system
to its original state once a decision change has been made.   Suicide is an
extreme example of an irreversible decision.   In other cases, the decision
can be reversed but only at large social or economic cost.   Humpty-Dumpty
is the literary personification of this very important objective of water
resource and many other civil systems.

Some decisions are completely irreversible but in a continuous sense.
That is, the state of the system, s , can be changed by arbitrarily small

increments over time, t, or space in one direction but it can not be reversed. Mathematically, this form can be represented by $\partial S/\partial t \geq 0$. We can burn fossil fuel but we cannot "unburn" it. Other decisions are completely irreversible in either direction. In some cases the irreversibility is a matter of degree (i.e. social and economic cost), either continuous or discontinuous. A highway is an excellent example of a variable "irreversibility" since it can be removed or expanded only at considerably greater cost than if the proper decision had been made originally.

10.3 SOURCES OF UNCERTAINTIES AND ERRORS IN MODELING

Not all of the uncertainties or risks involved in water resources systems analysis are associated with the system itself. A significant uncertainty, all too often ignored in the quest for quantitative predictive models, is that related to the degree to which the various models used actually represent the significant behavior of the real system being modeled. These uncertainties can be introduced through the model's topology, its parameters, and the data collection and processing techniques. Model uncertainties will often be introduced through human error of both commission and omission. An "optimized" decision set is truly optimal only if the mathematical model used to generate that decision set closely represents the significant behavior of the actual system over time and space. The fact that some of the socio-economic elements of the real system are capable of reacting competitively or complementarily to the decision-maker's choice of decision set only emphasizes this shortcoming of most mathematical models. In fact, there are actually no civil systems involving a single decision-maker, despite this customary assumption in optimal decision modeling.

The necessary condition for reasonable utility of any decision set obtained through optimization is that the important responses of the real system to those decisions are the same as those produced by the mathematical model within a tolerable error. Since water resources decisions are very often made only once, it may be difficult to evaluate modeling errors, let alone reduce them to quantitative probability measures. This significant source of uncertainty is probably one of the major reasons for the slow and cautious adoption in civil systems of the products of research, particularly systems analysis modeling. The validity of the optimal solution $\underline{x}^*$ to any maximization or minimization problem depends (among other things) on the accuracy with which the mathematical model represents the real system, In particular, this accuracy depends on the closeness to the real system of the model's input-output relationships. The sources of uncertainties and errors can be associated with the following six major categories of model characteristics:

  (i)   Model Topology - $(\underline{\alpha}_1)$

  (ii) Model Parameters - $(\underline{\alpha}_2)$

  (iii) Model Scope or Focus - $(\underline{\alpha}_3)$

  (iv) Data - $(\underline{\alpha}_4)$

  (v)   Optimization Technique - $(\underline{\alpha}_5)$

  (vi) Human Subjectivity - $(\underline{\alpha}_6)$

The above six categories are discussed hereafter in some detail.

10.3.1 Model Topology $(\underline{\alpha}_1)$

Model topology refers to the order, degree and form of the system of equations which represent the real system.   For example, a dynamic system might be represented by a system of differential equations (ordinary or partial); a static system might be represented by sets of algebraic equations such as polynomials, etc.

Consider for example, a groundwater system of both confined and unconfined aquifers. In order to model the dynamic response of the hydraulic head in the aquifer to any future demands (withdrawals or recharge) on the groundwater system, one may use a system of differential equations. Linear second order partial differential equations may be adequate for modeling the confined aquifer, whereas nonlinear second order partial differential equations (PDE) might be needed for the unconfined aquifer. Furthermore, a homogeneous aquifer may be adequately modeled by a two-dimensional system, but a stratified and non-homogeneous one ought to be modeled by a three-dimensional PDE, etc.   Clearly, in each case, a selection of one model topology over another introduces a source of uncertainties and errors in the accuracy of the model's representation.

Model topology is particularly important in decision making for optimization.   Almost any functional form can be used to approximate the absolute value of any cause-effect relationship.   However, optimal decisions are usually not as concerned with the magnitude of these functions as with their derivatives (or incremental ratios). Thus a linear least squares regression model of a basically non-linear response, because of the characteristics of linear system optimization, is very apt to select "decisions" at points which in fact have the greatest error in the representation of the true derivative.

10.3.2 Model Parameters ($\underline{\alpha}_2$)

Once the model topology has been selected, the choice of model parameters (often called parameter identification, parameter estimation, system
identification, model calibration, etc.) determines the accuracy with which
the system model represents the real system.

Consider the groundwater system discussed earlier. Once the customary system of parabolic partial differential equations is selected, the
proper values of the coefficients need to be determined (e.g. storativity
and transmissivity as functions of the spatial coordinates). This parameter
estimation (identification) process introduces a source of uncertainties
and errors in the accuracy of the calculated values of the parameters and
in turn in the model itself.

10.3.3 Model Scope ($\underline{\alpha}_3$)

Model scope refers to the type and level of resolution used in the
model for the description of the real system. Four major descriptions are
identified in water resources systems.[3] These are:

  (i) Temporal description.
  (ii) Physical-Hydrological description.
  (iii) Political-Geographical description.
  (iv) Goal or Functional description.

The above descriptions are discussed in some detail hereafter. The characteristic parameters of uncertainties and errors associated with the selection of the model scope is denoted by the set $\underline{\alpha}_3$.

In referring again to the ground water system, one may wish to study
the behavior (response) of the system under planned development for short,
intermediate and long-term planning horizons (temporal description). The
groundwater system itself, which may consist of several aquifers, may be
decomposed on the basis of the physical-hydrological characteristics or
political-geographical boundaries. Finally, if the groundwater system is
to be managed as part of a larger water resources system with concern for
water quality, storage, recharge, and so on, then different decompositions
may be more advantageous, such as goal description. Clearly, while these
four descriptions have individual merits, each describes the system from a
narrowed point of view. The system in totality may never be well-represen-
ted by any one description, and thus the selection of model's scope introduces yet another source of uncertainties and errors in the system's repre-

sentation. Scope is a particularly important factor where the system is
controlled by many relatively independent decision-makers, each with somewhat different objectives. Even so, such systems are often modeled as
though a single "rational" decision-maker was at the helm, i.e. as if a
single point of view can be asserted.

10.3.4 Data (α_4)

Access to sufficient representative data for model constructions,
calibration, identification, testing, validation and hopefully implementation, is obviously a very important element in systems analysis. Clearly a
lack of either accurate or sufficient data due to the collection, acquisi-
tion, processing, analysis, etc., may cause substantial errors in the results.

Consider again the above groundwater system: the value of the model
parameters determined through the identification procedure is likely to depend on the available data. An insufficient number of sampling sites, the
number of samples, and sampling accuracy (within each spatial location) may
introduce significant sources of uncertainties and errors in the system model.

10.3.5 Optimization Techniques (α_5)

Once the mathematical model has been constructed and its parameters
identified, the selection and application of suitable optimization methodologies (solution strategies) introduces another source of uncertainties and
errors in the solution derived from the system model.   In the groundwater
system discussed earlier, the selection of the method of numerical integration of the system of PDE with the associated grid size, boundary and initial conditions, computer storage capacity and accuracy, etc., all introduce sources of uncertainties and errors in the solution.   As another example, consider a nonlinear objective function with a nonlinear system of
inequality constraints representing the behavior of a power and water supply system.   If the optimization method applied for solving this system is
the simplex method (via linearization of the system model), then the accuracy of the solution obtained may be questionable.   This particularly is
true for highly nonlinear systems.

It is important to note that the selection of the optimization technique generally coincides (or should) with the model's construction. Consequently, any exchange between the sophistication (or simplification) of
the model and the accuracy (or approximation) of the solution should be

made at an earlier stage of the system's analysis.

10.3.6 Human Subjectivity (α_6)

Human subjectivity strongly influences the outcome of systems analyses in water resources (as well as in other areas). Human subjectivity may include:

    (i)   The background, training and experience of the analyst(s),

    (ii) Personal preference and self-interest, and

  (iii) Proficiency.

Clearly, human subjectivity can influence all of the other five major categories of model characteristics.

A civil engineer, a hydrologist or a systems engineer, for example, all addressing the problem of planning for the above ground water system development and predicting the waterhead response to withdrawals and recharges, may each conceive a different approach or methodology. While human subjectivity plays a very important role in the selection of all the major categories of model characteristics, each of which introduces sources of uncertainties and errors in the system model, no quantitative analysis is available in this respect.   Rather than attempt to quantify such cause and effect relationships here, the importance of each characteristic is indicated and a framework for their analysis is suggested.

In analyzing the sources of uncertainties and errors as they affect sensitivity, stability, irreversibility and ultimately optimality, the system's analyst may have available the following knowledge about the augmented vector $\underline{\alpha}$:

    (a) A complete knowledge of $\underline{\alpha}$ is available

          namely, $\underline{\alpha}$ is a deterministic variable.

    (b) The vector $\underline{\alpha}$ is a stochastic variable but an estimate

          of its probability distribution function is available.

    (c) The vector $\underline{\alpha}$ is a stochastic variable where no knowledge

          is available on the probability distribution function.

It is assumed that for any given system some analytical functions can be constructed relating sensitivity, stability and irreversibility to α.   Furthermore, depending on which element of $\underline{\alpha}$ is under consideration, the knowledge of its mean and variance can vary between (a) and (c). In any event, noncommensurable objective functions will result regardless of the degree of knowledge of $\underline{\alpha}$.   The Surrogate Worth Trade-off (SWT) Method can be utilized to solve this problem of noncommensurability among multiobjective functions.

10.4 FORMULATION OF RISK OBJECTIVES FOR WATER RESOURCE SYSTEMS

There probably is no standard approach to the specification of risk objectives in general, given the almost infinite possible number of combinations of system-modeling characteristics discussed in section 10.3.   To a large extent, each case may have to be treated de novo to assure that the modeling errors introduced by standardized approaches do not introduce more uncertainty than the risk element being analyzed.

The basic question is: Risk of what? It may be the risk of a reservoir going dry. In exactly the same physical situation, it may be the risk of failing to meet a minimum prescribed level of service, such as "firm water"or "firm energy". Thirdly, it may be the risk of divergence from the prescribed level of service.   In fact, all three risk elements may exist simultaneously.   The first two would constitute discrete units of the risk objective vector.   The latter constitutes an objective vector with an infinite set of components between zero and the prescribed level of service. For example, it may be desirable to know the risk of failing by 500 cubic feet per second (cfs) and by 1500 cfs, and it may be just as desirable to know the risk associated with 1000 cfs or any other point. In most cases, however, this continuous vector of objectives can be modeled at a small number of discrete points (e.g. 0, 500, 1000, 1500) and the intermediate points estimated by interpolation or curve fitting if the corresponding risk and optimal policies are relatively insensitive.

For example, let $F(x)$ be the expected net economic benefit of selecting a level of services of x. It is desired to set $F(x)$ as high as possible while minimizing the risk that the reservoir (of total capacity Q_{max}) and the stochastic inflow $y(t)$ will not be sufficient to provide minimum level of service at all times within the next n time periods. This is perhaps the simplest form of the risk problem, but it serves to illustrate a number of important characteristics which must be carefully considered.

Proceeding with this simple formulation, a knowledge of the statistical characteristics of the hydrology allows development of a large number of "equally likely" hydrographic sequences of n time periods each. Let today's reservoir level be q_0 units, and q_i be the storage at time period i; the water release at time period i, r_i, will be limited by the water inflow at period i, y_i, as well as the maximum capacity of the reservoir, Q_{max}. Any specified feasible release policy x, should satisfy the constraint: inflow - available storage space $\leq$ x . For time period i:

$$y_i - (Q_{max} - q_i) \leq x$$

The actual reservoir releases r_i might then be established for the policy x.

$$r_i = x \text{ if } y_i - (\Omega_{max} - q_i) \le x \le y_i + q_i$$

$$r_i = v_i - (\Omega_{max} - q_i) \text{ if } y_i + q_i > x_i$$

$$r_i = y_i + q_i \text{ if } y_i + q_i < x$$

Using this (or any other) fixed decision rule, the set of equally likely hydrographs can be used to determine the probability that $r_i < x$ at least once in any n period time horizon. In this way the probability of failure to meet minimum service levels at least once in n time periods is calculated as a function of service level. If this quantity is designated as $P_n(x)$, then the vector optimization problem is

$$\max_{x} \quad [1 - P_n(x), \ F(x)]$$

Subject to: constraints on input hydrology,
constraint on reservoir capacity, and
non-negative constraints on initial
reservoir conditions.

Since $P_n(x)$ and $F(x)$ are fundamentally different quantities, this is a vector optimization of noncommensurable functions and it can be treated using the Surrogate Worth Trade-off Method. Note that the optimum policy and acceptable risk levels will depend on the initial storage level chosen, hence this represents a family of optimizations.

There are several other representations of the risk element of this problem. For example, the risk objective can be defined as the probability that the decision level x will not result in a failure within n time periods (n = 1,2,3,...,N). In this case a probability distribution can be generated for $\hat{P}_n(x)$ for each level of x considered. The result is a family of optimization problems of the form

$$\max [1 - \hat{P}_n(x), \ f(x)] \qquad n = 1, 2, \ ..., \ N$$

Subject to constraints as before.

Once again the problem can be treated using the Surrogate Worth Trade-off Method.

At this point most readers will have probably wondered why we did not simply determine the probability of failure, assess an appropriate economic penalty function and proceed to maximize the mathematical expectation of the resulting single economic objective. This is a very valid question

and in certain circumstances it would be the correct decision model topology to follow.   The validity of this approach, however, depends upon the
skill and accuracy of constructing the penalty function for dropping below
a delivery of x in any time period.   In fact, this is what the Surrogate
Worth Trade-off Method does, except that instead of attempting to evaluate
the penalty (a very subjective matter in risk cases), attention is focused
on the simpler question whether the decision-maker is willing to accept a
specific (computable) increase in risk in order to obtain a specific (computable) increase in his benefit.   As was shown in Chapter 3, it is not
really necessary to know the answer to the latter question in an absolute
quantitative sense, but rather only in the ordinal (rank order) or qualitative sense of one being of greater value than the other. A penalty function
on the other hand must be numerically accurate over all possible values of
x, otherwise the derivatives on which optimization usually rests may be
badly in error.   If the proper penalty function can be accurately determined, both methods should lead to identical results.

The validity of using a penalty function and optimizing mathematical expectation is open to serious question in a number of counts in problems involving water resource systems.   To be valid the process must in
fact eliminate the basis for a residual risk objective, and this will be
possible only if certain conditions are met.   Two major types of problems
are discussed below: in the first type, the decision-maker must expect to
have a large number of applications of the decision, large enough so that
his actual experience with the decision can reasonably be expected to be an
adequate unbiased sample of the corresponding probability distributions. A
decision-maker who only gets one trial with its corresponding result, is
not readily consoled by the mathematical expectation. He still must consider separately whether the risk is worth the gain. Anyone would be willing
to make a series of 1,000,000 bets of $1.00 each on the black numbers on
the roulette table if he were paid even money for the two green house numbers as well as the black. However, very few would bet $1,000,000 with one
and only one bet allowable under exactly the same circumstances. The mathematical expectancy is exactly the same, but the relative desirability is
obviously quite different.

In addition to the requirement that an adequate number of experi-
ences are possible, the mathematical expectation must also be related to
other risk producing components of the system such as irreversibility, stability and responsivity.   For example, firm power contracts in the Central
Valley project of California require the level of firm power contracted to

drop to the lowest power output actually delivered, whenever power output falls below the "firm" contract level. This constitutes a discontinuity in the economic objective function itself, and mathematical expectation as normally defined is thus not adequate for the risk element concerned.

The type of risk-return problem described above is essentially one of risk produced by the sensitivity of the parameters describing the system.

A second type of risk problem is that produced by sensitivity to the decision. This type occurs whenever the decision variable cannot be precisely controlled but varies about the decision point, resulting in a corresponding variance in output. Once again, if the proper conditions are met and an appropriate penalty function can be determined over the range of variance, it is appropriate to optimize the mathematical expectation of setting the decision variable to its mean value. If these conditions are not met, the problem again becomes a multiobjective optimization.

On occasion, it may be very difficult to assess the risk _per se_ in other than qualitative or judgmental terms. However, knowing that risk is related to sensitivity will permit a multiobjective analysis involving sensitivity and return, where sensitivity substitutes for risk. To do so, this requires using a definition of sensitivity which reflects these qualitative and judgmental factors. For example, a probability distribution of the variation of the decision about its selected value $\bar{x}$ may not be known (or readily determinable), yet it may be possible to estimate a "Span of control" x_c, including the most likely significant variation in x. Note that x_c may be a function of $\bar{x}$ since the ability to control x may depend on its magnitude.

An interesting problem in sensitivity arises in the optimal construction sequencing problem[+]. The methodology developed there determines the optimal order of construction for N water supply projects, each having a specific fixed capacity of Q_i. Any given estimate of the future demand for service results in some specific order of construction, depending on the time pattern of this demand. However, the latter is an uncertainty usually obtained by extrapolation processes which, while sometimes valid for short periods into the future, become much more unreliable as the time span increases. If it is presumed that an equally likely future demand (or requirement) function can be estimated for a short time period on the basis of past trends (estimating the mean and standard deviation of the error from past variances), it is possible to generate a number of possible future demand functions by methods similar to those used to generate simulated hydrologic sequences. In the absence of any other knowledge we can assert that these are best possible estimates of a number of "equally likely" future demands.

Using such a set of equally likely future demand functions the optimal sequence of project construction for each simulated future can be computed. If they are equally likely, it is possible to determine the frequency with which any particular project would be constructed first. The rank order of this frequency represents a numerical measure of the sensitivity of the supply-use system to the initial project decision under the conditions of demand uncertainty.

10.5 MEASUREMENT OF RISK-RELATED CHARACTERISTICS

In section 2, four risk-related characteristics of water resources systems were identified and defined in a descriptive sense. However, in order to incorporate these characteristics into decision analysis, quantitative measures will be required for each, and none is presently available. The purpose of this section is to explore in a tentative way some potential measures of sensitivity leaving quantitative expression of responsivity , stability and irreversibility for future discussion.

In this discussion the measurement of sensitivity is approached from the point of view of risk and uncertainty rather than from that of its more abstract mathematical connotation. However, it will be useful to begin with the latter.

Let the various causative factors of risk and uncertainty entering into the modeling analysis be identified by α_1, α_2,...,α_n. From its general definition, one measure of sensitivity is the rate of change of any systems output, objective, or decision with respect to the factors α_j as well as the rate at which this rate itself is changing. Thus a vector measure of the sensitivity of systems outputs y_i at a point could be expressed as:

$$ S_y = \left|\left| \frac{\partial y_i}{\partial \alpha_j} \right|\right| , \quad \left|\left| \frac{\partial^2 y_i}{\partial \alpha_j^2} \right|\right| , \quad \cdots , \quad \left|\left| \frac{\partial^k y_i}{\partial \alpha_j^k} \right|\right| $$

where $\left|\left| \cdot \right|\right|$ denotes a norm (the least squares as an example). In any practical application only the first few terms of this vector of derivatives would be significant.

Similar expressions can be written for the system objectives $f_i(\underline{x},\underline{\alpha})$, and for the "optimal" policy vector x_i^*. There are also sensitivities of each of the above with respect to each other, i.e., system objective may be sensitive to risks or uncertainties, to the system output y_i, or to the decision policy x_i^* and vice versa. A similar definition can be made with respect to any or all components of the constraint vector $\underline{g}(\underline{x},\underline{\alpha})$. Once a sensitivity norm is defined, the systems analyst may seek to minimize this norm, along with minimizing the overall system objectives $f_i(\underline{x},\underline{\alpha})$. This clearly lends itself to noncommensurable vector minimization problems where the surrogate worth trade-off method can be applied for its solution.

The "span of control" may have an important influence on the selection of the proper norm for the model's sensitivity. Consider Figure 10.2 where the graph is given of three functions denoted by Case I, II, and III, all of which possess the same maximum at α^*. Clearly, each of the three functions has a different sensitivity to α. Accordingly, a different norm

based on $\dfrac{\partial f}{\partial \alpha}$, $\dfrac{\partial^2 f}{\partial \alpha^2}$, or $\dfrac{\partial^3 f}{\partial \alpha^3}$ may prove to be advantageous in Cases I, II, and

III respectively. Note that the span of control of Case I, for example, is much wider than that of Case II or Case III.

Unfortunately the use of various orders of derivatives as a measure of sensitivity has the serious failing of being valid only within the immediate neighborhood of the decisioned point x_i^* and its associated output y_i and objective f_i. However the risk or uncertainty factor α_i may cause the actual x_i, y_i and/or f_i to deviate substantially from the decisioned values x_i^*, y_i and/or f_i^*. An irrigator may "decide" to apply 3 inches of water at an irrigation but his ability to control that decision at exactly 3 inches leaves much to be desired. His actual irrigation may be anywhere from two to four inches with corresponding variations in output (soil moisture availability) or objectives (profit on the crop).

10.6 SUMMARY AND CONCLUSIONS

In this chapter a number of questions associated with risk and uncertainty have been tentatively explored for the purpose of stimulating further analysis and research into the quantifications of these factors for use in multi-objective optimization analysis. A great many problems exist in water resources systems and other civil systems involving resources in which avoidance of risk and uncertainty are often in fact the dominating objective. If suitable quantitative measures of these objectives can be formulated then the surrogate worth trade-off method can be used to determine the optimal or at least superior combinations of risk and various forms of return.

Direct measures of risk to be avoided can be defined in certain situations. Example of hydrologic risk quantification (to be minimized)

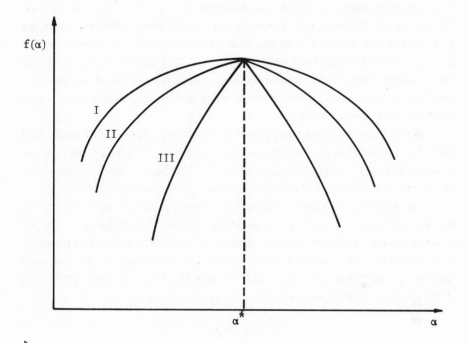

Figure 10.2   Variable Span of Control

were developed as an example of treating risk due to chance controlled non-decisioned inputs or system parameters.

More complex risk and uncertainty situations develop when decisions cannot be made with precision control but rather will vary about the decision values in some random or quasi random manner. When the number of repetitions of that decision is also small so that mathematical expectations may be meaningless to the decision-maker, an important form of risk is introduced.

Such imprecision of control may introduce risk or uncertainty through several systems characteristics. Four such system characteristics were identified and descriptively defined: sensitivity, responsivity, stability and irreversibility. In addition several types of modeling errors were identified which can lead to imprecise control, imprecise predictions of the real response or both, hence having equivalent ability to create or accentuate risk and uncertainty.

Because they are somewhat singularly related to the specific systems concerned, generalizations on responsivity, stability and irreversibility are not discussed in this more general discussion. Sensitivity on the other hand would appear to be amenable to more generalized quantifications as discussed in the previous section. Each of the above measures is useful under successively more general circumstances ranging from systems controllable within close limits to those which can be only approximately controlled within broad limits.

This somewhat preliminary analysis and discussion indicates that quantitative measures of risk can be defined and utilized as objectives to be optimized in a multi-objective control. In some instances even uncertainty (no probability distribution data) can be treated adequately.

An indication, however, is not an accomplished fact and much insight and analysis will be required to quantify the major risk factors involved in common water resources systems adequately to allow their inclusions in multi-objective decision analysis. Because of the singular and sometimes overriding importance of this issue it is hoped that this discussion will stimulate that insight and analysis.

FOOTNOTES

1. The gambler's ruin problem is described by Hall and Dracup, [1970].
2. These estimates are described by Nadeau, [1950].
3. The development of the descriptions for water resources systems can be found in Haimes and Macko, [1973].
4. The optimal construction sequencing problem is solved by Butcher, Haimes and Hall [1969].

References

1. Butcher, W. S., Y. Y. Haimes, and W. A. Hall, "Dynamic Programming for the Optimal Sequencing of Water Supply Project," Water Resources Research, vol. 5, no. 6, p. 1196, 1969.

2. Hall, W. A. and J. A. Dracup, Water Resources Systems Engineering, McGraw-Hill Book Co., New York, 1970.

3. Nadeau, R. A., The Water Seekers, Doubleday and Company, Inc., New York, 1950.

4. Haimes Y. Y. and D. Macko, "Hierarchical Structures in Water Resources Management", IEEE-Systems, Man, and Cybernetics, vol. SMC-3, no. 4, pp. 396-402, 1973.

Chapter 11

EPILOGUE

This chapter summarizes the characteristics and advantages of the SWT method and indicates further extensions of the method.

Multiple objective planning and decision making is an important problem for most civil systems. The various objectives involved generally cannot be represented in common units, hence to find the best policy, the decision maker must make a mental analysis of the trade-offs that might be achieved. The SWT method combines these same mental processes with mathematical analysis in a procedure which provides a converging systematic approach. The mathematical analysis applies only to the quantitative functions. High precision functions are not diluted by low precision functions since each is a separate and distinct vector throughout. The decision-maker need respond only to his sense of satisfaction of levels of objective attained and his sense of desirability of simple "ΔA" vs. "ΔB" trade-off possibilities. This is exactly what a decision-maker always does or at least tries to do. The analysis assumes that those are the best trades possible at that level of attainment, where sub-optimal combinations are not presented. Thus the SWT method models a process very similar to the real decision process where a single decision-maker is involved, yet it substantially reduces the number of combinations of levels of objectives attained that need be considered.

11.1 ADVANTAGES OF THE SWT METHOD

The major characteristics and advantages of the surrogate worth trade-off method are:

(i) Non-commensurable objective functions can be handled quantitatively.

(ii) The surrogate worth functions, which relate the decision maker's preferences to the non-inferior solutions through the trade-off functions, can be constructed in the functional space and only later be transformed into the decision space.

(iii) The decision-maker interacts with the mathematical model at a general and a very moderate level. He makes decisions on his subjective preference in the functional space (more familiar and meaningful to him) rather than in the decision space. This is particularly important since the dimensionality of the decision space is generally much larger than the dimensionality of the functional space.

(iv) The SWT method provides the decision maker with additional quantitative information on the non-inferior (pareto
optimum) space. In particular, the trade-off functions

$(\lambda_{ij} = -\dfrac{\partial f_i}{\partial f_j}$, $i \neq j$, $i,j = 1,2,...,n)$, which are the

slopes of the non-inferior curves in the functional space,
are of significant importance to the decision-maker by
providing the relative trade-offs at any level of objective
achievement between any two objective functions.

(v) Computational feasibility and tractability have been
demonstrated through the solution of several example
problems.

(vi) The applicability of multiobjective analysis via the
surrogate worth trade-off method to several problems
in water resources planning--water quality maintenance, reservoir operation and construction, etc.-- has
been demonstrated.

(vii) The availability of operational methodologies, such as
the SWT method, encourages and enhances the systems
modeling and pattern of thinking in multiobjective functional terms. Thus more realistic analyses may result
by eliminating the need for a single objective function
formulation.

11.2   FURTHER DEVELOPMENT OF SWT METHOD

While the SWT method can be utilized to advantage in most multiobjective optimization problems there is still a number of areas in which substantial improvement in effectiveness should be possible.

One of the implicit assumptions in the development of the SWT method
is that there is a single quantitative value which can be assigned as the
DM's assessment of preference.   In most real problems involving civil systems, there is a number of decision-makers who will have diverse opinions,
hence there can be several such quantitative values. This problem was discussed briefly in section 3.4 . With further refinements, the application
to multiobjective optimization involving multiple decision-makers should be
feasible.

In addition, refinements may be useful for determining the types of
questions to ask decision-makers. The requirement that the increment of

each objective be small, must be counter-balanced by the requirement that it must still be large enough for the DM to be able to perceive the differences. Thus far, a method for determining the optimum size of these increments for all problems has not been found. Guidelines to improve the consistency and accuracy of decision-makers will also aid in applying the SWT method to real problems. The sensitivity analyses suggested in the algorithms of chapters three through six should also be utilized when applications to real problems are considered.

One of the advantages of the SWT method is that almost any algorithm for determining non-inferior points can be incorporated into the procedure, so that as new and improved algorithms are developed, the SWT approach can be constantly updated. The only requirement is that the trade-off rates $\lambda_{ij} = - \partial f_i / \partial f_j$ must be determinable as part of the solution.

This book has considered only deterministic problems; however, there is a large amount of work on decision theory for problems with stochastic decisions, objectives and constraints. Adapting the SWT approach to these problems would increase its usefulness. In particular, the SWT method can treat significant factors of risk and uncertainty as objectives.

The dynamic problems in this book considered only objectives which were presumed not to change with time, (e.g. integrals over time). The SWT method could be extended to dynamic problems whose objectives are time dependent-- where the value of some function at each point in time is important $(f_i(t) = L_i(\underline{x}(t),\underline{u}(t),t)$ for $i = 1,2,\ldots,n)$ or where the integral up to that point is important over the entire trajectory $(f_i(t)= \int_0^t L_i(\underline{x}(s),\underline{u}(s),s)$ ds for $i = i,2,\ldots,n)$. For example such problems may arise in long-term planning problems where the capital cost as well as the operations, maintenance and replacement cost is important. Problems of this form will require great modification since the trade-off rates λ_{ij} and the worth functions W_{ij} will also be functions of time.

Another area of application is to use the surrogate worth trade-off method as a coordinator in multilevel hierarchical models. Since the submodels in multilevel models often have noncommensurable and competing objectives, a coordinating analysis which can handle multiple objective problems is useful.

Finally, the most important application is to implement the SWT method in problems with real decision-makers.

AUTHOR INDEX